EAI/Springer Innovations in Communication and Computing

Editor's Note

The impact of information technologies is creating a new world yet not fully understood. The extent and speed of economic, life style and social changes already perceived in everyday life is hard to estimate without understanding the technological driving forces behind it. This series presents contributed volumes featuring the latest research and development in the various information engineering technologies that play a key role in this process.

The range of topics, focusing primarily on communications and computing engineering include, but are not limited to, wireless networks; mobile communication; design and learning; gaming; interaction; e-health and pervasive healthcare; energy management; smart grids; internet of things; cognitive radio networks; computation; cloud computing; ubiquitous connectivity, and in mode general smart living, smart cities, Internet of Things and more. The series publishes a combination of expanded papers selected from hosted and sponsored European Alliance for Innovation (EAI) conferences that present cutting edge, global research as well as provide new perspectives on traditional related engineering fields. This content, complemented with open calls for contribution of book titles and individual chapters, together maintain Springer's and EAI's high standards of academic excellence. The audience for the books consists of researchers, industry professionals, advanced level students as well as practitioners in related fields of activity include information and communication specialists, security experts, economists, urban planners, doctors, and in general representatives in all those walks of life affected ad contributing to the information revolution.

About EAI

EAI is a grassroots member organization initiated through cooperation between businesses, public, private and government organizations to address the global challenges of Europe's future competitiveness and link the European Research community with its counterparts around the globe. EAI reaches out to hundreds of thousands of individual subscribers on all continents and collaborates with an institutional member base including Fortune 500 companies, government organizations, and educational institutions, provide a free research and innovation platform.

Through its open free membership model EAI promotes a new research and innovation culture based on collaboration, connectivity and recognition of excellence by community.

More information about this series at http://www.springer.com/series/15427

Melody Moh • Kanta Prasad Sharma
Rashmi Agrawal • Vicente Garcia Diaz
Editors

Smart IoT for Research and Industry

Springer

EAI
RESEARCH MEETS INNOVATION

Editors
Melody Moh
Department of Computer Science
San Jose State University
San Jose, CA, USA

Rashmi Agrawal
Manav Rachna International University
Faridabad, Haryana, India

Kanta Prasad Sharma
GL Bajaj Group of Institutions Mathura
Mathura, Uttar Pradesh, India

Vicente Garcia Diaz
University of Oviedo
Oviedo, Asturias, Spain

ISSN 2522-8595 ISSN 2522-8609 (electronic)
EAI/Springer Innovations in Communication and Computing
ISBN 978-3-030-71484-0 ISBN 978-3-030-71485-7 (eBook)
https://doi.org/10.1007/978-3-030-71485-7

This Springer imprint is published by the registered company Springer Nature Switzerland AG
The registered company address is: Gewerbestrasse 11, 6330 Cham, Switzerland

Preface

Internet of Things (IoT) refers to physical and virtual objects that have unique identities and are connected to the Internet to provide intelligent applications that make energy, logistics, industrial control, retail, agriculture, and many other domains "smarter." Internet of Things is an evolution of the Internet that is rapidly gathering momentum driven by the advancements in sensor networks, mobile devices, wireless communications, networking, and cloud technologies. Experts forecast that by the year 2050, there will be a total of 50 billion devices/things connected to the Internet.

This book is written as a textbook on the Internet of Things for educational and research programs at colleges and universities, and also for IoT vendors and service providers who may be interested in offering a broader perspective of the Internet of Things to accompany their own customer and developer training programs.

The typical reader is expected to have completed a couple of courses in programming using traditional high-level languages at the college level, and is either a senior or a beginning graduate student in one of the science, technology, engineering, or mathematics (STEM) fields. We have tried to write a comprehensive book that transfers knowledge through an immersive "hands on" approach, where the reader is provided the necessary guidance and knowledge to develop working code for real-world IoT applications. Concurrent development of practical applications that accompanies traditional instructional material within the book further enhances the learning process, in our opinion.

The book comprises a total of 13 chapters. We introduce the reader to advanced topics on IoT including IoT data analytics and tools for IoT, fog technology for IoT, and IoT-based search engine in agriculture. Review studies on collecting and analyzing data generated by the Internet of Things in the clouds are described for multidimensional research. Through generous use of hundreds of figures and tested code results samples, we have attempted to provide a rigorous "no hype" guide to the Internet of Things. It is expected that diligent readers of this book can use these exercises to develop their own IoT applications or tools. We adopted an informal approach to describing well-known concepts primarily because these topics are covered well in existing textbooks, and our focus instead is on getting the reader firmly on track to developing research-oriented robust IoT applications as opposed to more hypothesis. The opinions in this book are those of the authors alone.

Chapter 1: "Applications of Blockchain in Various Domains" provides an overview of the enthusiasm for blockchain innovation of transactions made in cryptocurrency, and this chapter leads a conscious plan, concentrating on the sole purpose of collecting all the important developments on blockchain innovation. The primary focus is to comprehend the momentum research points, identifying difficulties in the existing system

Chapter 2: "IoT-Based Unique Air and Noise Pollution Monitoring System" describes the cause of pollution and extreme level of issues arising from the same with help of IoT-based systems, right from being an active contribution to the global warming crisis and to causing localized health problems to people who breathe the same air.

Chapter 3: "New Frontiers in Managing and Controlling Industrial Processes Through IoT" describes current businesses in the pandemic era that are highly affected due to low availability of manpower conditions and managing through IoT-based models for machine-to-machine communication to enhance IoT-based innovations in health, agriculture, home appliances, research, and communication as an intelligent machine.

Chapter 4: "Smart Self-Immolation Prediction Techniques" highlights the factors responsible for suicidal tendency, but also performs a study on several models and machine learning–based applications for anticipating the suicidal tendency of a person by remote monitoring, which is very essential to prevent the attempt of such a devastating crime. Priority has been imposed upon depression as a cause of suicidal tendencies.

Chapter 5: "A Review of Particle Swarm Optimization in Cloud Computing" describes the ever-growing demand for online computational resources, which is mandatory, and the best resource allocation algorithm to allocate the resources to its end users. This chapter also focuses on the comprehension analysis of swarm- and cloud-based algorithms founded on heuristics, AI, and ML concepts.

Chapter 6: "Role of Satellites in Agriculture" introduces an immense development in the field of agriculture with one of the major areas of sensor technology. The images from satellites have proved to be a boon in precision agriculture and have increased the efficiency of production as well as aided farmers with the help of machine learning and big data vision to make the right decision at the real time.

Chapter 7: "IoT in Green Engineering Transformation for Smart Cities" provides the concept of IoT in green engineering transformation for smart cities and role of IoT in establishing communication among devices to implement smart living and smart environment. It further discusses the role of IoT in different sectors of a smart city, along with technologies used for green IoT for research directions and challenges in the field of green engineering.

Chapter 8: "A Study on Optimal Framework with Fog Computing for Smart City" introduces the issue and challenges of mobile users for increasing rapidly and having high requisition of location-based and localized information. This motivates and encourages the use of fog computing. Fog computing can be renamed as edge computing, it is an extension of cloud computing. This study proposes fog computing techniques and models for minimizing the overall latency while placing data on the fog, in the current scenario, to reduce the delay tradeoff when the task is offloading.

Chapter 9: "Changing World: Smart Homes Review and Future" describes smart homes principles and associated technologies, building block concepts with inter connectivity through sensor technology, and communication protocols.

Chapter 10: "Providing Security and Managing Quality Through Machine Learning Techniques for Image Processing Model in Industrial Internet of Things" Describes The proposed model is efficient to computerize structure of the CPU lines of manufacture in an industry. Machine learning techniques for suitable classification for abnormalities and data security aspects on the cloud.

Chapter 11: "Solar-Powered Portable Charger Using IoT-Based Technique" introduced with portable charges, electrical devices without charges with external energy. The concept of solar energy technologies for emitted by the sun and convert electricity for clean and renewable power sources and applications.

Chapter 12: "CBC (Cipher Block Chaining) Based Authenticated Encryption for Securing Sensor Data in Smart Homes" introduces well-equipped automation with various sensors. This chapter provides an authenticated encryption mechanism for secure communication and original message ensuring data policy and mechanism.

San Jose, CA, USA — Melody Moh
Mathura, Uttar Pradesh, India — Kanta Prasad Sharma
Faridabad, Haryana, India — Rashmi Agrawal
Oviedo, Asturias, Spain — Vicente Garcia-Diaz

Contents

Editors' Biography

Melody Moh obtained her BS in electrical engineering from National Taiwan University, and MS and PhD in computer science from the University of California, Davis. She joined San Jose State University in 1993, and currently serves as the chair with the Department of Computer Science. Her research interests include IoT, cloud computing, mobile networking, information security, and machine learning applications. Her research has been funded by both government and industry; she has published over 160 refereed papers in international journals, conference proceedings, and books, and has consulted for various companies.

Kanta Prasad Sharma has a rich innovative teaching experience of more than 13 years. Dr. Sharma has a PhD in information technology from Amity University and master's degree from UPTU Lucknow, India. At present, he is working as assistant professor of computer Science and engineering, and coordinator of the R&D cell at GLBAJAJ Group of Institutions. He has published 16+ research papers with SCI & SCOPUS indexing in international journals and conference proceedings. Dr. Sharma is a member of ACM, Computer Society of India, and IEEE as well as governing body member of ERDA and IAENG (International Association of Engineering). He has been engaged as faculty guide for more than 15 undergraduate and graduate students and also supervising five MTech research scholars in the field of IoT, navigation and communication, wireless sensor networks, and finger control technology. Dr. Sharma is a member of the editorial board of more than 5 international and national journals. He is presently an editor at Springer and Willy publications and TPC member of more than seven international conferences.

Rashmi Agrawal, PhD, has 19+ years of experience in teaching and research, and is working as a professor in Department of Computer Applications, Manav Rachna International Institute of Research and Studies, Faridabad, India. She has qualified the UGC-NET. Dr. Agrawal is associated with various professional bodies in different capacities. She is life member of the Computer Society of India; senior member of IEEE, ACM, and CSTA; and senior member of Science and Engineering Institute (SCIEI). She is book series editor of Innovations in Big Data and Machine Learning (CRC Press, Taylor and Francis group, USA) and Advances in Cybersecurity (Wiley). Dr. Agrawal has authored/co-authored many research papers in peer-reviewed national/international journals and conferences which are SCI/SCIE/ESCI/SCOPUS indexed. She has also edited/authored books with national/international publishers (Springer, Elsevier, IGI Global, Apple Academic Press, and CRC Press) and contributed chapters in books. Currently, she is guiding PhD scholars in sentiment analysis, educational data mining, Internet of Things, brain–computer interface, and natural language processing. Dr. Agrawal is associate editor of the *Journal of Engineering and Applied Sciences* and managing editor of the *Journal of Sciences and Technology*.

Vicente García-Díaz is an associate professor in the Department of Computer Science at the University of Oviedo. He is a software engineer with a PhD in computer science. Dr. García-Díaz has a master's degree in occupational risk prevention and the qualification of university expert in blockchain application development. He is also part of the editorial and advisory board of several journals and has been editor of several special issues in books and journals. Dr. García-Díaz has supervised 100+ academic projects and published 100+ research papers in journals, conferences, and books. His teaching interests are primarily in the design and analysis of algorithms and the design of domain-specific languages. His current research interests include decision support systems, health informatics, and e-learning.

Chapter 1
Applications of Blockchain in Various Domains

Jishnu Bhardwaj, Raunak Negi, Preeti Nagrath, and Mamta Mittal

1.1 Introduction

Blockchain [1], mostly referred to as the foundation technology behind Bitcoin [2], is one of the emerging automations in the market luring a lot of recognition from firms, startups, and the media. Blockchain has the possibility to alter numerous industries and make procedures more democratic, safe, clear, and effective.

With huge amounts of data getting processed every day due to digitization of documents, it becomes very essential for every institution to efficiently command the security menaces and acquire significant cost efficiencies. Now this is the point where blockchain technology, with its promise of redistributed possession and unchangeable and cryptographic [3] security of evidence, is catching the eye of the C-suite directors. Financial people are the first ones to take advantage of this technology even though it is still at an early stage. A recent study published by the World Economic Forum [1] has predicted that banks and governments around the globe are at ease to examine and experiment numerous blockchain templates.

With as many as 90 central banks [4] involved in blockchain discussions all over the globe, around 2500 patents [4] have been filed over the last few years and around 80% of the banks have started the Distributed Ledger Technology (DLT) projects [5]. The blockchain technology is on its way to become the new dominant force in the world.

In this chapter, we explicate the uses and applications of blockchain [2] technology.

J. Bhardwaj (✉) · R. Negi
Department of Computer Science, G.B. Pant Government Engineering College, New Delhi, India

P. Nagrath · M. Mittal
Department of Computer Science, Bharati Vidyapeeth's College of Engineering, New Delhi, India
e-mail: Preeti.nagrath@bharatividyapeeth.edu

M. Moh et al. (eds.), *Smart IoT for Research and Industry*, EAI/Springer Innovations in Communication and Computing,
https://doi.org/10.1007/978-3-030-71485-7_1

1.2 Related Works

Blockchain technology is finding its applications in almost all-important areas of the industry, and hence it has been a center of attraction for a lot of R&D. Over the last few years, it has been observed that the blockchain ecosystem has been booming, and there have been a lot of key concepts in the field of blockchain, dispersive applications, and digital identity-based financial exchanges. The previous and ongoing researches focus on the key concepts of the aforementioned, such as decentralized applications, Ethereum, smart contracts, blockchain-based identities, and dispersive autonomous organizations, and the other applications are discussed below.

1.3 Introduction to Blockchain

1.3.1 Blocks and Hashes

"Blocks" are the digital containers in which blockchain transactions are stored. Each block that has been created is linked to its parent block via hashes which are also referred to as unique digital fingerprints. These hashes are imprinted in the form of a header at the topmost part of each block. The history of all the transactions that are contained on the blocks is then linked back to the "genesis" block. The data stored in these blocks are completely secure and cannot be tampered with even by the person who has created these blocks. This procedure is done by the non-dependent nodes that come to a dispersed agreement for every transaction that has been made [6] (Fig. 1.1).

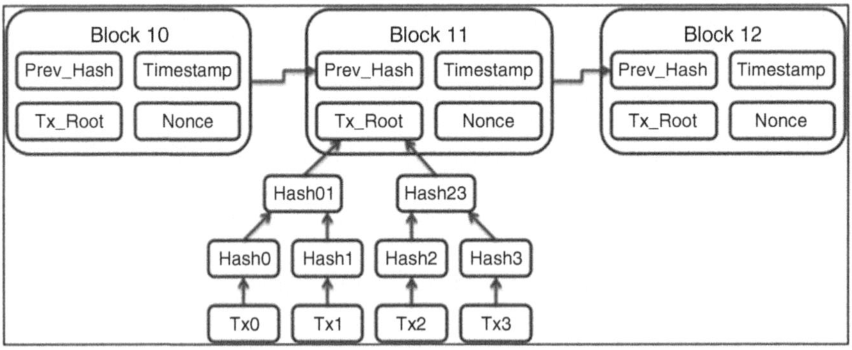

Fig. 1.1 Blockchain hashing algorithm [7]

1.3.2 *Keys, Tokens, and Transactions*

The prime concern of blockchain phenomenon is the strategy of ownership and its capacity of allocation of digital tokens to other users. The credit of this accomplishment is through an expertise called "public key cryptography." Here two keys are available – public and private keys. These keys are actually preserved in a convenient database called wallet. A publically exchangeable address for the user depicted by a unique string of numbers and letters is generated by public key. The public key is signed by a private key to produce a unique digital signature. Transaction on the network is done by the submission of this signature. The units of this account are the digital tokens and are used in keeping a track of blockchain [8]. Rules created by designers inspire different blockchain systems to have separate supply of tokens that function in accordance to its designs. Globally, companies charge for doing business on different digital tokens – either with each other or with governmental currencies. The innovative discovery of digital value exchange is its idea of reflection of two primary sources of weaknesses or susceptibility in traditional means of value exchanges:

1. Requiring an outside organization to confirm payments, verify contracts, and authorize settlements involving more than two parties [9]
2. Allowing different, and potentially conflicting, transaction records to be stored by the transacting parties and allowing for each record to be subject to alteration

The above mentioned problems/events are taken care by blockchain protocols and protected by advanced cryptographic features that are sometimes not valid through earlier protocols. Thus, prolonged decentralization and the use of distributed ledgers have affected advanced consensus mechanisms and created a huge system of availability, efficiency, and competency [10] (Fig. 1.2).

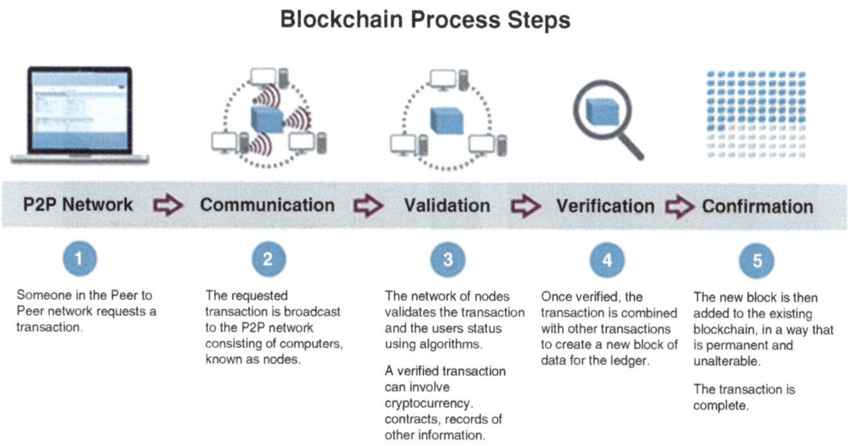

Fig. 1.2 Step-by-step transaction process [1]

1.4 Bitcoin: Introduction

The three largest cryptocurrencies in today's global market are Bitcoin, Ripple, and Ethereum. These cryptocurrencies are also the largest blockchain systems, since blockchain is the primary technology that allows them to function [11]. In this chapter, we will talk about the largest cryptocurrency of the three, that is, Bitcoin. The initial researches on Bitcoin began in the early 2008 and the first breakthrough was the publication of the white paper "Bitcoin: A Peer to-Peer Electronic Cash System" by an unknown person under the alias Satoshi Nakamoto [12]. He describes Bitcoin as the first "peer-to-peer version of digital cash" that allows payments to be sent in a cryptographic format from one person/organization to the other without the fear of it being hacked. More importantly, Nakamoto's idea represented the initial solution to the "Byzantine General's Problem" [13]. It was considered as a respected digital puzzle that was believed to be unsolvable before Bitcoin came into existence. It revolves around the fact that a group of non-trusted independent actors need to verify a given scenario.

Bitcoin is easily able to solve this issue by introducing the concept of "Proof-of-Work" (POW) chain which makes it necessary for third-party mediators to verify themselves and prove that they are solving difficult problems in a small amount of time. Computational power of each node determines the proof of work of the network. Transactions between two parties are verified without the intervention of a third party. In simpler words we can say that POW is a complex, expensive, and time-consuming task to incorporate but easy for others to verify. To understand this better let us use an example.

Previously it was stated that hashing the document consists of performing a daunting operation to the numbers, such as adding or multiplying. As long as the nodes on the network all verify, there is an agreement. If billions of nodes on the network show up, and one node produces the hash, that particular node would be altered as per the harmonious mechanism. So, it is next to impossible to tell what the original numbers were [14] (Fig. 1.3).

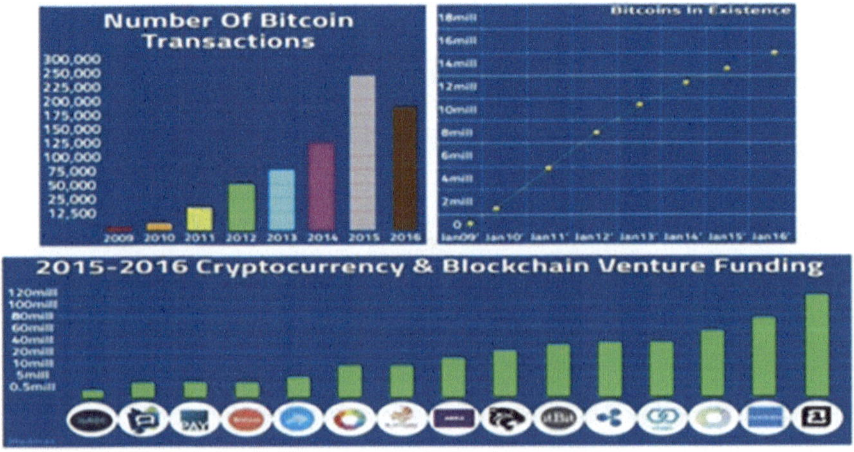

Fig. 1.3 Graphical representation of transactions made using Bitcoin [7]

1.5 Blockchain in Finance Services

In today's world, blockchain technology is of a very good use in looking after finance arrangements for the banks. Under this technology, all the important files such as a goods shipping bill and tax tally can be recorded in a concentrated depository where all the concerned parties involved can control the data in real time.

One great example of this is the Ethereum blockchain [15] which sanctions more comprehensive and secure business channels and well-organized processes and lessened costs in trading and finance. It enables digitalized [1] trade to take place within a less time period, at no extra cost, with greater levels of adaptation.

Its advantages are as follows:

- *Dependability*: Its varied planning removes the points of single failure and also gets rid of the need for data mediators such as transfer representatives and communication system employees. It also ensures to enable secure [2] application code which has been designated to be hack-proof against venomous third parties – which makes it nearly impossible to hack.
- *Lucidity*: It adopts high standards and agreements acting as the only shared origin of truth for system participants.
- *Reliability*: Its dependable archives make the job simpler for various parties in a business channel to cooperate and direct data.
- *Programmability*: It helps in the creation and implementation of computerized contracts – hack-proof, digital software [3] that helps the business logic – creating increased reliability and effectiveness.
- *Seclusion*: It provides tools that are leading in the market for granular data privacy across all the available softwares, allowing careful sharing of data in business communication channels. This dramatically improves dependability, reliability, and effectiveness while maintaining confidentiality.
- *Performance*: Its private channels are maneuvered to sustain many financial agreements per second and a time to time flow in network activity.
- *Extensibility*: It acts as a mediator between private and public chains [1] and offers each enterprise the global reach and high coherence.

1.5.1 Ethereum

Ethereum is a dispersive platform that allows us to implement extra Distributed Applications or DApps [7]. The DApps are written with smart contracts. The implementation of one or more intelligent contracts can form a DApp. An intelligent contract will run exactly as planned, without any pause, restriction, tampering, and interference by third parties.

Figure 1.4 represents the structure of Ethereum blockchain.

One of the most significant advantages of using Ethereum blockchain to run smart contracts is that it allows the interaction between two and more smart

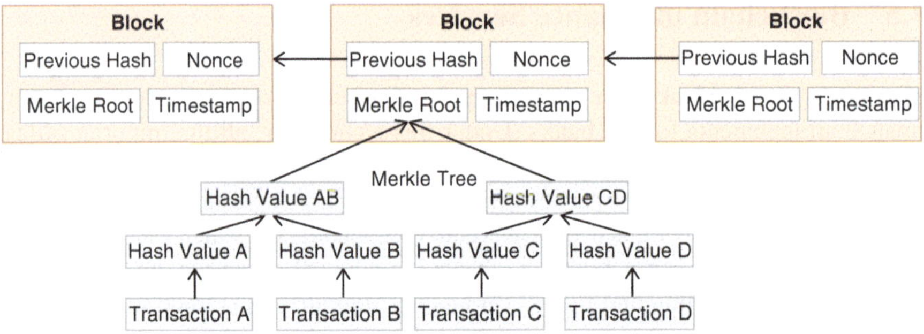

Fig. 1.4 Ethereum block structure. (From "Blockchain Platform for Industrial Internet of Things," Arshdeep Bahga, Vijay Madisetti, 2016)

Fig. 1.5 Comparison between Ethereum and Bitcoin. (From CoinMarketCap, *The Economist*)

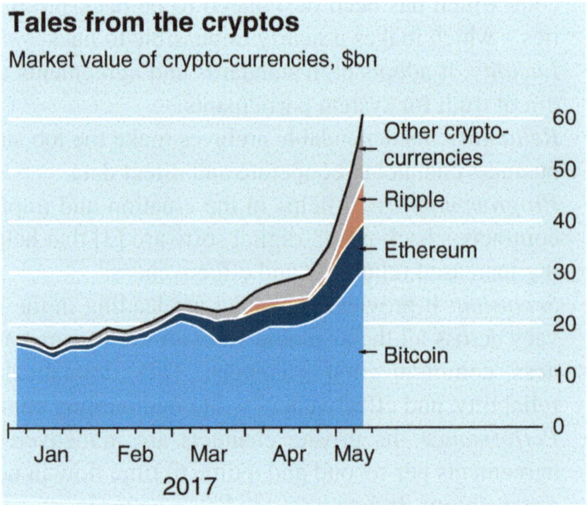

contracts. Integrating the agreement protocol and other things is already taken cared of beforehand, only the logic of the application needs to be written and executed, DApps cannot be created with Ethereum blockchain, and only those types of DApp can be created whose functions are supported by Ethereum. The internal currency of Ethereum is called Ether. To implement intelligent contracts, Ether is needed. The total market value observed for different types of cryptocurrencies along with Ethereum until May 2017 is presented in Fig. 1.5. Ethereum's total market value is about $35 billion as represented in the figure.

1.5.2 *Smart Contract*

A smart contract is a dependable entity as it carries out tasks accurately without any possibility of downtime, restrictions, trickery, and intrusion by third parties. A contract contains variables, functions, function modifiers, events, structures, and enumerations. They also support inheritance. Inheritance is implemented in a contract by copying the code at compile time. Some smart contracts also support polymorphism, i.e., “Serpent” or “solidity” is used to write a smart contract. Once the code is written, either Mist or Ethereum wallet (which was developed by the Ethereum Foundation) can be used to deploy it in blockchain. These smart contracts can store data, send and verify transactions, and interact with other contracts.

1.5.3 *Decentralized Autonomous Organization*

Blockchain technology along with Ethereum smart contract can together be deployed to unleash an immense variety of decentralized applications or DApps, such as a decentralized storage system, exclusive cryptocurrency, and Decentralized Autonomous Organization (DAO) [7]. Availability of exclusive version of cryptocurrency can essentially be deployed to address a variety of the internal functionalities of different companies. More than 30 decentralized human-less venture capital platforms are specially provided by DAO for investors who wish to invest. The investment is decentralized and distributed. Figure 1.6 describes the role of DAOs in the decentralized ecosystem. The DAO acts as a link between the blockchain, the

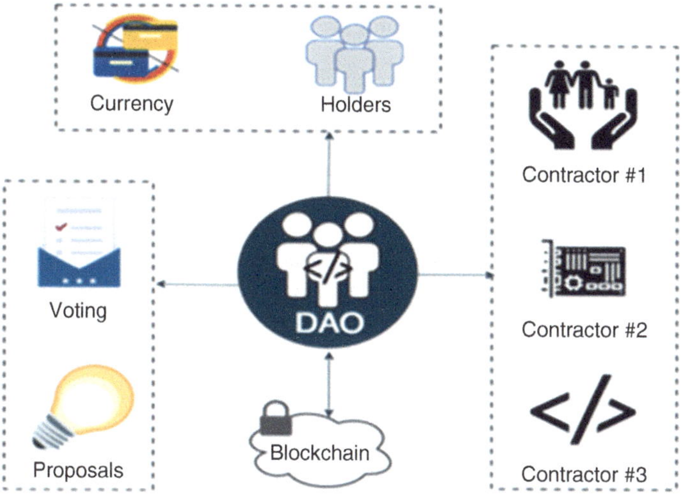

Fig. 1.6 Role of DAO in a decentralized system. (From https://blog.codecentric.de/en/2017/09/decentralized-autonomous-organization-blockchain/)

currency holders, the administrators with voting rights and proposals, and the contractors, as shown pictorially in the figure below.

Efficient hardware utilization is achieved using a file storage which has a decentralized structure, which promotes users to rent unused or stagnant data drives. WeiFund and uPort are examples of such platforms. WeiFund promotes crowdfunding in Ethereum and uses smart contracts to establish an open platform. uPort provides for users to exercise total control over their identity, and no institutions or authorities are involved.

1.6 Application of Financial Trade Services

1.6.1 Central Bank Digital Currency [16]

A Central Bank Digital Currency (CBDC) [16] is the digital form of money kept in a central bank, which is the legal compassionate created and backed by a central bank. CBDC is managed through a centralized archive [5], thus tripling the security of transactions between banks and individuals. A recent study that was conducted by the bank stated that more than 70% of organizations are actively developing proofs of concept for the CBDCs. Aspects that define a Central Bank Digital Currency are as follows:

- *Digital aid*: CBDCs are considered to be centralized digital assets and are considered as the only source of truth.
- *Central bank backed*: CBDC represents the public's claims legally.
- *Central bank operated*: The supply of CBDC is fully controlled by the central bank.

Some features of this digital currency are as follows: [17]

- *Trust*: A blockchain-based CBDC system enables central banks to control the flow of currency and at the same time protect the privacy of the CBDC's use to the end users.
- *Programmability*: CBDC is able to restrict third-party access to the information. This feature comes in handy while protecting the user's data.
- *Availability*: Distributed systems such as blockchains ensure that data is made available to its users in addition to developing trust and transparency around transactions. Ethereum has proven to be a good example in this case.
- *New Developments*: A blockchain-based CBDC benefits from the creative products and services that are being developed across the blockchain ecosystem, including unguarded wallets, no knowledge of cryptography, and dispersed transactions (Fig. 1.7).

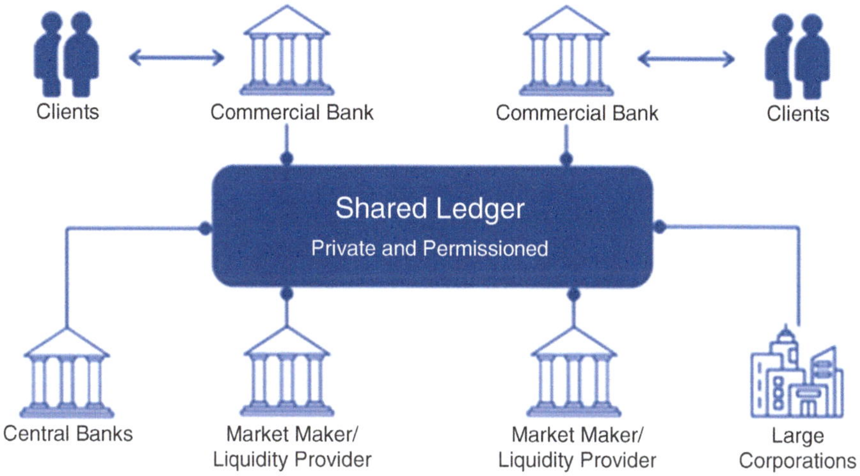

Fig. 1.7 Central Bank Digital Currency [18]]

1.6.2 Codefi Payments [19]

Codefi Payments offers a suite of blockchain-based payment solutions designed to make global transactions faster and more secure. It allows the businesses to easily accept cryptocurrency payments and also supports the commonly adopted payment structures that include invoicing, checkouts, etc.

1.6.2.1 Payment Plans

The product is registered with one or more subscription plans.

1.6.2.2 Automated Smart Contracts

Payments are accepted and tested using smart contracts directly from the Codefi Payments interface.

1.6.2.3 Integrate with the Software

A person needs to choose from three integration options:

Sending a payment link or email to the customers directing them to a Codefi-hosted payment page.

Install the Codefi pay widget and create a custom interface using API.

1.6.3 Trade Finance

Trade finance is the funding of the international business chain. The finance industry thrives on paper-based procedure which is more exposed to security threats. Single negotiations can take up to 90–120 days in order to process the letter of acknowledgment, substantiate documents, and demonstrate reliance among collaborators. Blockchain can digitize the entire trade finance cycle with increased security and effectiveness. It enables more clear administration, decreased clarifying times, low money requirements and decreased risks of trickery, human mistakes, and counterparty risk [20].

It allows the following:

- Digitized and genuine attestation and KYC/AML data with instantaneous verification of financial assets
- Digitization of assets enables quick settlements
- Production of more effective financial structures through secure systems and digitized processes
- Production of a harmonious financing vehicle for the entire trade cycle, which eliminates the legal practice of dealing with independent finance vehicles for each stage of the trade [21]

1.7 Blockchain in Healthcare

There has been a thorough and comprehensive study of blockchain ever since its introduction via Bitcoin [1]. It mainly focuses on the nonmonetary or nonfinancial cases, but it has been observed that healthcare and its services is one of the few industries on which blockchain is anticipated to have serious effects and consequences.

Research in this sector has been fairly recent but is rapidly increasing, leading to health informatics and researchers in the path of struggling, having difficulty to keep up with the speed of research progress in this sector. The approach here is based on the PRISMA policy [22] and guidelines along with an organized mapping study, where well-thought-out search protocols are used to explore scientific databases, for determining, extracting, and evaluating every relevant and suitable publication. Numerous studies have shown different applications of blockchain in the healthcare sector, but there is a shortage of sufficient model execution and research to describe, identify, and interpret the potency of these described cases. Growth of blockchain applications in healthcare, limitations they pose, and possible areas of growth have been discussed briefly in this section. Hence, more research is required in order to better comprehend, describe, and evaluate the usefulness of blockchain in the healthcare industry (Fig. 1.8).

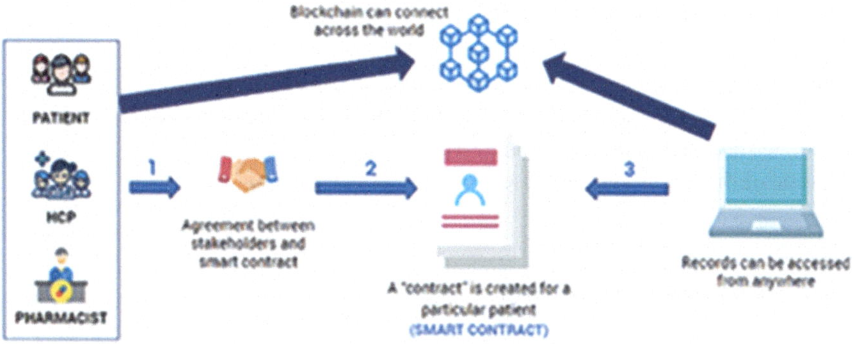

Fig. 1.8 Data flow in blockchain-powered healthcare systems [23]

Taking our first step in methodical mapping study, we circumscribe the given research questions in order to unscramble the cutting-edge research on blockchain technology in healthcare.

1.7.1 Applications and Real-World Cases of Blockchain

The essential question here is to grasp various parts and sections of healthcare that blockchain has been implementing. By reviewing and screening the relative articles from scientific databases, recognition of healthcare problems of blockchain with their possible solutions can be determined, which can be followed by isolating and dividing those problems that are resolved and settled better using other approaches, techniques, and means. An outline of problem territory in healthcare where blockchains are applicable helps the researchers, analysts, and professionals to emphasize their interests and theories to those bright areas of this technology's application.

1.7.2 Prepared Blockchain Models Based on Identified Real-World Applications

In the fields of scientific works and literature, there are several areas of application of blockchain. However, when talking about working prototypes, not all of these suggestions have been rendered and executed. It is, hence, important to comprehend the degree of usage scenarios of implementing blockchain-based healthcare applications. This helps us draw our attention to the sections and areas in which there exist research holes and the need to transform research focus to these affected zones and sections.

1.8 Applications in Healthcare

After screening and preliminary examination of some of the papers, one or more than one situation of some of the identified block use cases was addressed by the papers, we decided to discuss them [23].

Some of the major areas of discussion are as follows.

1.8.1 Electronic Medical Records

One popular real-world application of blockchain in the health department is the managing of Electronic Medical Records [24], abbreviated for EMR. EMRs, used exchangeably with Personal Health Records "PHRs" [25] or Electronic Health Records "EHRs" [26], are associated with electronic formation, warehousing, and administration of the user's (patient's) health, personal, or medical-related data.

The major properties and characteristics of blockchain, which include decentralization, constancy, dependability, sturdiness, data protection and secrecy, make it a hallmark for stocking, managing, and administrating patient's Electronic Medical Records or "EMR."

The European General Data Protection Regulation or "GDPR" [27] is a regulation in European Union, which forbids the access of delicate data of patients they explicitly give their permission. Blockchain is widely proposed as a doable technology too, since it is consistent with "GDPR" and gives the users full control and authority over their data.

1.8.2 Pharmaceutical Supply

Talking particularly about the pharmaceutical industry, another real-world application of blockchain is in health supply chain management [28]. Since delivery and supply of fraudulent and forged medicines can have ghastly conclusions for the patient, it is still a usual challenge faced by the drug industry.

To address this problem, blockchain has been found having the capability. There are some companies that are working detecting prescription drug deception via blockchain, as mentioned by Engelhardt [29], in his examination. Nuco and HealthChainRx are some of the mentioned companies [29]. The concept is recording all the transactions which are related to the drug prescriptions on a blockchain network, where all the interested parties such as producers, distributors, doctors, patients, etc are linked. In case of any modification or malicious adjustment of the prescription by stakeholder parties, it can be identified.

1.8.3 Biomedical Research and Education

In the case of the biomedical department, blockchain has a very intriguing application. Elimination of false data and exclusion of undesirable results in the research can be achieved by blockchain. The anonymity of users in blockchain, which is its inherent property, helps patients grant the authorization and approval for their data to be utilized for clinical studies and experiments. Moreover, the immutable property guarantees the data integrity.

This clear and open nature of blockchain technology makes it easier to replicate and reproduce research from the blockchain data. This is why it is expected that blockchain would surely revolutionize the biomedical research department.

1.8.4 Health Data Analytics or "HDA"

Blockchain technology offers an opportunity to utilize and exploit the emerging technologies. Accomplishing predictive analytics of data and progressing research in medicine is now achievable by powerful techniques of deep learning and transfer learning [30]. A research, conducted by Juneja and Marefat, looks for ways in which blockchain is used in a deep-learning architecture for arrhythmia classification [31].

1.8.5 Protecting and Ensuring Patient Data

One of the most important qualities and aspects of using blockchain technology in the health department is to keep our medical data safe and protected. Security has been a major issue in this industry. Between 2009 and 2017, more than 176 million patient records have been exposed in data infringement and leaks [32, 33]. The violators and hackers stole important information like credit card and banking details, along with health and genomic testing records.

Blockchain, with its ability and power to preserve a trustworthy, decentralized, and transparent log of patient data, proves itself as a powerful technology for security cases. In order to protect the delicate and sensitive medical data, blockchain masks out the identity of the individual user via complicated and secure codes, making it a powerful combination of being transparent and private. This noncentralized kind of the technology enables healthcare service providers, doctors, and patients to exchange the information efficiently and safely (Fig. 1.9).

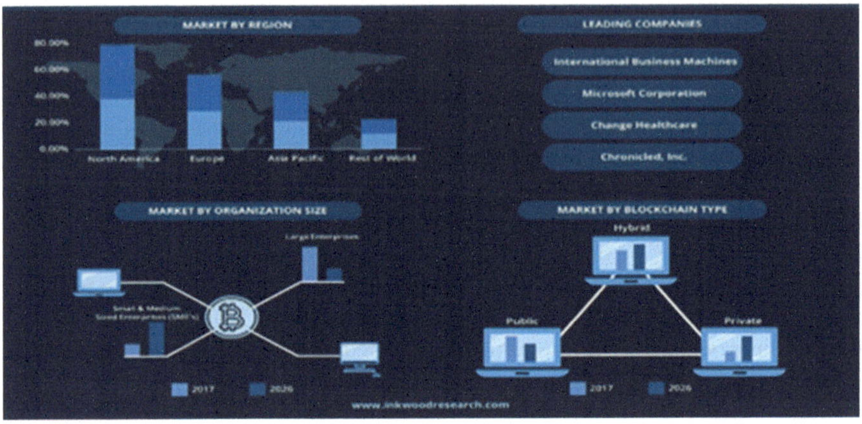

Fig. 1.9 Statistical forecast of blockchain in healthcare market [32]

1.9 Blockchain in Architecture

In the current scenario, construction engineering faces countless challenges and problems in regard to confidence, information sharing, and automation process to name some. We know blockchain is a decentralized operation and information management technology, gathering more and more interest from academic as well as industrial aspects [34]. Despite that, a large portion of the current research, practices, and usages are concentrated on blockchain itself (technical issues and limitations) or its applications in financial service sectors like Bitcoin [1] as discussed above.

The possibility and capacity of blockchain applications in the construction sector is our main aim in this section. Looking at the possibilities, the three types of applications that are enabled by blockchain are as follows:

1. Enhancing current processes of contract management
2. Management of supply chain
3. Leasing of equipments

The construction industry has been regarded as one of the world's most disconnected and high impact sectors. The infrastructure and framework projects all over the globe have highly fragmented, diffused, and sophisticated supply networks. Crossrail project [35], London, is a great example to represent this phenomenon.

In order to handle such an expanded supply chain, monitor of work in progress, schedule and register costs and payments, colossal action, and resources are necessary. Not limited to this, construction projects encounter various types of mistakes, blunders, postponement, and crashes at various levels and degrees. The lack of responsibility and accountability in this business has been a problem for ages, along with highly tight profit margins. Firms find ways to cut costs and try to dodge and shift blames, faults, and charges from the consequent failures. This is where blockchain can help us to make such procedures more competent, transparent, productive, and accountable.

Enhancing project management, providing a reliable database, computerizing of processes and documentation, and acting as a record book for legislative or regulatory concerns are some of the examples of main use cases of blockchain in the construction and building industry. Some of the possible blockchain applications have already been introduced and had an effect on the economy. A couple of these applications can be either straight away applied to the construction industry or serve as a basis for custom applications in capital construction [36, 37].

1.10 Applications in Architecture

1.10.1 Automated Ledger for Subcontractors

The usual step for acquisition of the building starts with a client, then its consultants, and the main contractor. The main contractor further undertakes subcontractors for carrying out all the specialized works, which generally vary from masonry to carpentry to painting. Several brokers and middlemen sit between the client and the final outcome.

Tracking of these middlemen and subordinates, with their tasks, has proven to be a hurdle. Regular checking and tracking of the amount of materials and equipment required, the clearance of access to the building, the deadlines, and checking of working hours of labor and other masonries are all tedious tasks.

An "automated ledger" [38] can potentially serve as a way to overcome all the obstacles mentioned above. This will make it easier to keep track of the construction and also to identify reliable subcontractors for a project, representing a subject-matter-of-the-contract relationship.

1.10.2 "Smart Contracts" to Initiate Milestones

One of blockchain's most well-known applications, "smart contract" [39], is being extended to different industries including construction. Automation of contractual operation and paperwork can be done by putting smart contracts in place, which saves money, frees up valuable resources, and makes project delivery efficient.

These contracts can be established at different levels and layers of the construction meshwork, from management of the building to the payment process. Clients and managers are able to track building processes and identify the people accountable for these processes because the contract is only activated when a task, set beforehand, is completed. It can also be used as a milestone-based payment network, which will generate computerized agreements, hence creating an verifiable record of operations.

1.10.3 Decentralizing Automated Organization

Groups of "smart contracts" can be utilized for creating a DAO [40], or "Decentralizing Automated Organization." It is a system which is run via rules encrypted as computer programs utilizing smart contracts. The integration of blockchain and the Building Automation System [41] (BAS) allows building's DAO, for example, to place an order for something like new fan fitting, accept the delivery and take responsibility for it, get some individual for its installation, and pay the supplier as well as the installer.

This DOA's wallet, which is linked to the wallets of those individuals who live in the building, is used for making payments. Collection of rents, corporate and enterprise fees, insurance installments, etc., are all within the reach and autonomously managed by the company's DAO.

Besides the code, the DAOs do not have a hierarchical framework and arrangement. For its construction phase, it requires human input to set the conditions and makes decisions to meet those requirements, which can include light fitting, paint color, maximum speed, etc. These requirements are met and decisions are made by starting a series of if/then that uses bundles of interlinked smart contracts carried out between customer and different members of the project personnel, main contractor, etc. [42]. DAOs are open source, hence transparent and incorruptible, in theory at least.

1.10.4 Life Cycle Ledger

In case of applications going beyond smart contracts, blockchain technology is particularly useful. It has the ability to work as a ledger to preserve a report of the construction process from one extreme to the other and record the blockchain information which can be cited back to when needed. For enterprise servicing, renovation, and regulatory conformity, it proves to be especially useful.

In a similar way, the ledger can store warranties and validations. Due to its nature, it can protect the construction process from tampering and fraud. This is because trials and outcomes can be stored, trailed, and traced on the blockchain, enabling easy comparison to building codes and requirements, along with streamlining audits throughout and following the construction period. In the post-construction phase, ledgers can be used for keeping records of developments in the life cycle of the building. It is a potent tool for enhancing operations and optimizing institute capacity to pinpoint and exchange (occasionally huge amounts) data with the individuals and agencies.

1.10.5 The Management of Data

Blockchain as a service can be of special importance, in situations where the technology provides a communication podium between different users, enabling them to manage the performance of their tasks in an efficient manner. It has the ability to permit tracing of data, boost decision-making procedures, and cut down the waiting time in transmission [43, 44].

1.11 Voting

Recently, many countries have begun using electronic voting systems or EVS. The first attempt in the world was made by Estonia [45] for voting in its national elections [46]. The voting was done via the Internet and a national identification card. This card is utilized for authentication, encryption, and signatures. Using the given electronic ID for verification, the voter downloads the voting application, and if eligible, a series of candidates would appear on user's screen and the vote was casted. This was followed by Switzerland adopting the same strategy for its state polls. In order to compete against conventional and customary ballot systems, the electronic voting had to have some support by being safe, reliable, and anonymous in nature. The security in this system had to be tight and impervious to protect the voters from any external interference and prevent any alteration or recasting of the given vote. Tor is one such browser, often used for deep-web services, to hide the security and implement the anonymity of the voters.

However, there are several drawbacks:

1. Politics in manufacturing: Taking into account the given political scenario, companies which produce these electronic voting machines could have manufactured them to "favor" the current ruling party. High scrutiny and quality checks will be required to overcome this.
2. Accuracy in recording intended votes: In case of breaking or misalignment of the touchscreen voting systems while its transportation or loading/unloading, the machines might misread or misinterpret the voter's purpose. For example, a user who wanted to vote for candidate B might vote for candidate A due to described problems. Such discrepancies on a large scale can lead to unfortunate and terrible outcomes.
3. Fraud: In all the major countries, there are countless claims and proclamations of fraud by the losing parties, in different levels and increments. Millions of votes can be easily nullified or falsified, if the voting machine created and dispersed by one vendor to different polling booths turns out to be dishonest and malicious in nature [47].
4. Hacking: The machine manufacturers and election jurisdictions usually claim to not broadcast and relay the poll results from different precincts via the Internet,

but this does not stop them from transmitting results via direct LAN connection, via VPN [48] (Virtual Private Network) or by other similar means.

Hence, in order to overcome all these drawbacks, the idea in BEV, or "Blockchain-Enabled Voting," is simple. Explaining this structure in terms of current currency analogy, each user or "voter" is given a one vote or "coin." The credentials and other related information of the user are stored in the "wallet." When a user votes for a certain candidate, this coin is transferred from the user's wallet to the candidate's wallet. The coin is issued only once to prevent repetition of votes. However, the users are given the choice to modify their casted vote before a defined deadline [49].

Therefore, we can imagine that this way blockchain addresses two of the most universal concerns in voting today: voter access and voter fraud.

By using a smartphone or desktop/laptop, blockchain provides an encoded key and immutable personal IDs, which entitles voters to vote anonymously. A mobile-based Boston startup, Voatz [50], is an e-voting platform which employs instant verification and biometrics. An everlasting, unbending record is established when the public ledger binds an individual user, or in this case, voter, to each ballot. With the peer-to-peer network, no one can engage in nefarious activities, and if they do so, they are corrected or evident on the ledger. Hackers would need to hack the blocks where the transaction records are stored, before the introduction of any new blocks.

In order to provide permanence and prevent illegal or false addition, all the given votes are controlled, verified, and monitored by blockchain. The mobile phone version of this startup was tested during events such as student government elections, church group, non-profit organization, union voting, and subnational political party events. For town meetings in Massachusetts, this system was also used. Russia is another example of blockchain voting. The city of Moscow's Active Citizen program [51] was launched in 2014 and has more than two million users (Fig. 1.10).

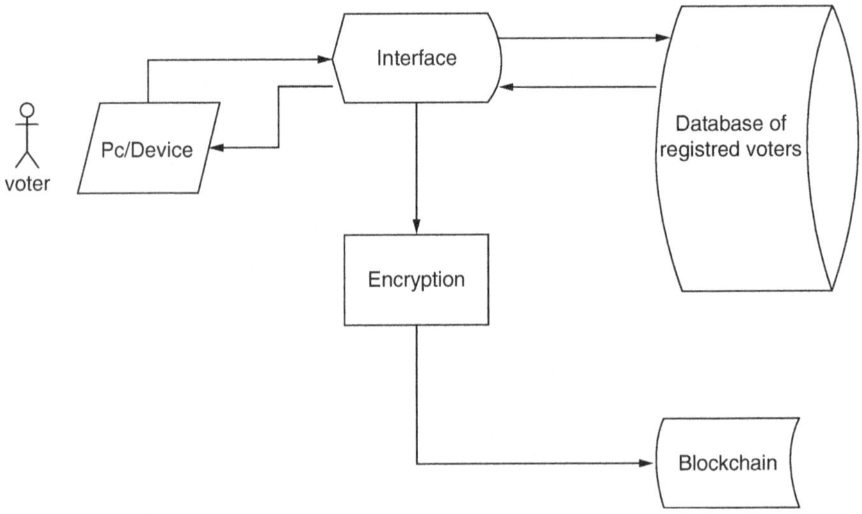

Fig. 1.10 Representation of the e-voting system [51, 52]

1.11.1 Opportunities and Benefits

1. The voting records generated by blockchains are encrypted in nature in order to address falsification of votes. The manipulation of the votes becomes impossible, since they are recorded securely, transparently, and permanently. This allows open scrutiny and public review of the votes while maintaining participant's anonymity. Even though it is said that nothing is secure, hacking into these records is nearly impossible and unfeasible.
2. Participation and access to voting is likely to increase with the help of improved identity verification by BEV. BEV is shown to improve the situation, as Voatz can accept the state IDs, passports, driver license, and seven other official documents for authorization and verification of the identity of voter [50].
3. Wiping out of any ambiguities is another benefit with BEV. With greater focus on promoting greater transparency and clarity to voters, it helps in reducing any likelihood of ambiguities. For some, the current online voting procedure might seem complex. Moreover, it is difficult to figure out whether the casted vote was intended or it was counted when casted. As we have discussed above, the public and open nature of blockchain makes it auditable.
4. Processing time of the current machines is high. Since the voting machines are in different areas at the time of polling, gathering them together is a difficult and time-consuming task. This reduces the efficiency of the counting, leading to monetary issues. Introduction of blockchain can metamorphose the entire process and time consumed. The outcome of the polls can be found instantly, after the voter finishes off voting. There would not be any need for long lines and manual labor of collecting the machines.

1.11.2 Challenges

The above discussion would sound wonderful, but only in theory. In order to implement the given ideas, many hurdles need to be conquered by the institutions and stakeholders. Public confidence in the security and accuracy of this technology is another key aspect for success. The complex blockchain's model and procedures might be a hindrance to the mainstream public. Moreover, the skills of the digital user is also a concern.

The political leaders might block the introduction of blockchain, as it takes away the central authority of the electoral commissions and agencies. Transparent being in the fiber of blockchain might cause the power shift from the authorities to the public, which in the current scenario is what they will try to stop.

The software and the technology is not up to the mark in order to efficiently implement the encryption and chaining. Since encryption is an extensive method, it is time-consuming for general smartphones and computers. Feasibility and power consumption will be the prime areas to tackle in terms of hardware shortcomings.

Estimates have suggested that, on average, there are 25–50 defects per 1000 LOC. For Ethereum [53], a blockchain-based platform for distribution, used by the Active Citizen program of Moscow, the numbers are nearly double. With the scale of hacking in modern technology, it has been found that Ethereum contracts are a "piece of cake" for them [54]. Moreover, sufficient observations have not yet been collected and processed to establish the scalability and extension of blockchain-enabled voting.

There have not been enough ledger technology and blockchain-based working models to accurately judge this technology in regard to present polling systems [55]. Full and complete execution of BEV for a domestic election has not yet occurred, though it has the ability to transform the voting in the coming future.

Politics-related violence during, before, and after elections is common in Africa and other developing countries [56]. The safety and transparency, with the reduction of electoral violence, can be ensured by BEV. More accurate and honest results can be predicted. All the voting-related expenditures would decrease, as it does not involve management and maintenance from any central authority. Finally, reduction in cost of paper-based elections and increment in participation of voters is expected after total implementation of this technology.

1.12 Music

The music industry consists of many entities, ranging from artists, labels, or record companies to retailers and streaming digital service providers [57]. The birth of streaming services has given a major boost to the music industry. There has been a shift from ownership to access, and the conventional revenue route has been disrupted. Many major record labels have witnessed huge drops in sales.

Blockchain has a massive potential to alter the supply chain, even though it is still in its early days. Open Music Initiative [58], OMI, is one of the many organizations which tries to bring the industry together to clarify and elucidate the benefits of investing in this technology. It assures to return the power back to creators via a fair distribution network. Freedom for accessing the transactional information and being paid more effectively is the idea. Introduction to original and creative models by the stakeholders or creators themselves is another potential opened up via this network. This would reduce the role of brokers and middlemen, like labels and publishers, and they might have to adapt themselves in the new environment according to the power balance.

Blockchain being transparent in its process allows the musicians and creators to see exactly where their money went and how much they are owed. With the current scenario, the artists are the last person to see how much profit they have made, even though they are the first one to put any work into it. Many of them are in the dark on how their profits are calculated, in which direction the money is flowing, and how people are listening to their music. It also makes it clear who the original copyright holder is, who was involved with their content.

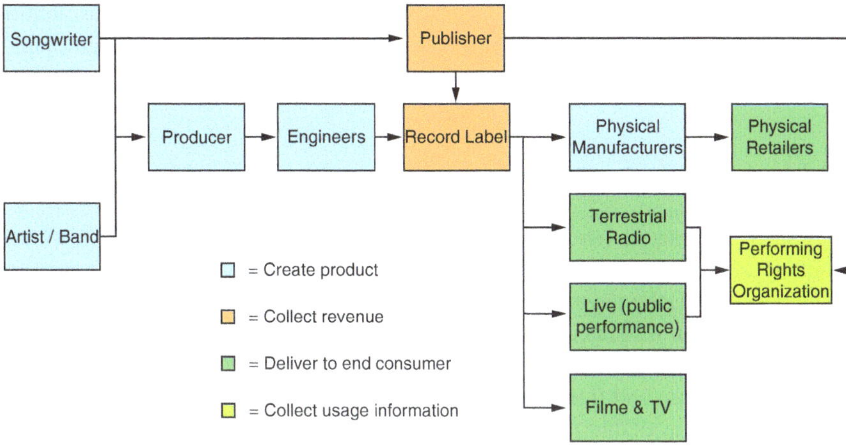

Fig. 1.11 Music supply chain before digital media [59]

No matter how excessive and bold these claims may seem to be, the power of blockchain can indeed cause a storm in the music industry. The major transformation it brings is removal of royalty payments to the middlemen and intermediaries. The establishment of a peer-to-peer network gives most of the power to the creators itself, and they are able to get the fair remuneration which they deserve (Fig. 1.11).

1.12.1 Applications

1.12.1.1 Maintaining Record

For the production of every song, the team behind it consists of lyricists, producers, engineers, and other members of the technical staff. Identification and payment process of these stakeholders for their IP usage is currently out of date. Imogen Heap [60], a London-born English songwriter, has stated, "One of the biggest problems in the industry right now is that there's no verified global registry of music creatives and their works. Attempts to build one have failed to the tune of millions of dollars over the years [...]. This has become a real issue, as evidenced by the $150 million class action lawsuit that Spotify is currently wrestling with."

Just like WAV and MP3 formats, dot blockchain or ".bc" [61] was presented as a dynamic format for music files. Its main purpose was to manage the rights of the media around the world. Data like the ownership rights, payment rights, and related information was to be incorporated inside this file format. Once the creation is done, it is made available to the users after adding it to the network. Dot Blockchain Media (now known as Verifi Media) is an organization whose goal is to deliver and provide the arrangement and structure to the artists, songwriters, melodists, audio engineers. and people in the music industry and its related fields, through open-source rules and licenses.

1.12.1.2 Smart Contracts

Using smart contracts can be game-changing to the musical industry, because of the automatic compensation to the creators and artists. Whenever a user streams or downloads a song, the payment generated is automatically sent to the stakeholder, whose information is located within these chains of network. Hence, no third-party app is required. Instant and spontaneous payment will be received by the artists as soon as they release and license their song. The information regarding the transactions will also be available for auditing and cross-processing. This information can be used for further analysis and in the field of "Data Mining" [62].

It is believed that this process is not suitable for high-frequency low-cost transactions, which is the case of streaming and music. But there are platforms that are using this method, charging the user's with cryptocurrency. The first track to generate and automatically circulate payments through smart contracts for all the involved parties was Tiny Human, composed by Imogen Heap. This song was accessible in Mycelia, which is run via Ethereum network. The currency used for transactions was Ether, which is an Ethereum cryptocurrency [63]. Even though the revenue generated at that time was not much, because of the unpopularity of blockchain applications in the general public, the fans were keen on trying out new things. This process displayed high potential in today's market.

1.12.1.3 Analysis and Model Innovation

With the information being available publicly, not only transparency increases but also allows the artists and creators to utilize this information to generate better and more robust business models. The demographics and geographic reach, preferences of the purchase of users, etc. can be easily studied and accessed by means of data mining. One such example is the popular streaming company Netflix [64], which examines and reviews millions of data generated real time by the users. This is one of the key reasons for the great success of the company. This comes under the "big-data" [65] analytics.

1.12.1.4 Revenue Management

Music, like many other industries, is a stockable good. The price of any song is variable, depending on the number of times it was streamed or downloaded in a day, the region-wise distribution it achieves, streams per person, etc. Such fluctuations and swings can vary the song value and usually increase the revenue captured by the creators.

For example, after examining the records, if one track is played more in an identified amount of time, then its price can be increased for that period of time and little less in other intervals. This allows users to either stream the song in the discount period, or other users that are not affected can still stream it at the costlier period. It helps to increase the range of the time when the song is played. Also, cuts and discounts could be given to those users who stream the given music for a given number of times. The theory of super-members and fans can be created, which would allow the users of such groups to pay less or not even pay after a given number of times.

1.12.2 Current State

Many organizations and companies can be found investing in research and the required platforms that enable appropriate and suitable functionality of blockchain. The Open Music Initiative (OMI), one of the discussed organizations, is in the works of preparing a protocol for music rights of holders and creators. The purpose is to create an API system in order to support other interested parties for designing and improvement of their own system, rather than building a database to provide a collection of such records.

Other agencies like Bittunes, Peertracks, and Voise are creating indigenous systems to allow their users to download and stream music via blockchain. For royalty payments and related stuff, companies like Revelator and Blokur are providing services in order to increase the profit and efficiency by data analysis. The given list of platforms are themselves providing streaming and downloading services, though they are in beta version.

- Musicoin
- This is a blockchain platform that enables customers to stream ad-free music from a list of standalone artists. Promising no middlemen, transparent contracts, and equitable remuneration, the platform uses Universal Basic Income or UBI model to guarantee fair compensation according to their contribution. Musicoin cryptocurrency is used for transaction of money and payments through smart contracts. Moreover, it is encouraged to tip the musicians through musicoin. The platform has plans to distribute merchandise and subscriptions to the super fans to increase the popularity and power of currency, and to connect better with its audience.
- Resonate
- Resonate is a music streaming platform with bold claims of being more affordable to its listeners in comparison to direct competitors like Spotify and Apple Music. The cooperative has a number of independent artists and labels, and its goal is to increase its popularity with the market and the users. Smart contracts are used to pay the musicians, and upon joining, musicians are paid directly through smart contracts and, consumers, as new subscribers, are given 3 hours of free streaming, after which they follow the pay-per-listen model.

- Ujo Music
- Like the discussed platforms, Ujo platform also uses cryptocurrency and smart contracts to pay their artists in accordance with "paid per song." The platform uses Ethereum for their payment system and provides free and paid streaming services.

1.13 Businesses in Blockchain

Businesses can grasp blockchains in different ways so as to gain an advantage over their rivals. They can modernize their core, eliminate extra costs, and make psychological property ownership and transactions crystal clear and automated. Companies consider applying blockchain technology in four aspects:

1.13.1 Accounting Settlement and Crowdfunding

Bitcoins can help businesses solve their problems which are related to funding. For instance, cryptocurrencies were created to fund companies who wish to implement digital payments and accounting settlements. The automation of electronic transaction management enhances the level of control of financial business execution, both internally and externally. In addition to this, blockchain technology represents an upcoming source of capital crowdfunding. The investors of major enterprises can obtain alternative entrepreneurial finance via token sales or initial coin offerings. Companies would be able to handle finance-related issues more flexibly by dealing in digital currencies that are based on blockchain technology.

1.13.2 Information Sharing

Data is the most valuable resource and it plays a vital role in every enterprise. Blockchain provides reliable storage facilities and efficient usage of data. As a dispersed and secure ledger, this technology can be used to tackle digital assets for several companies in different capacities. Decentralized data storage refers to the fact that data is not given to a centralized agency but instead to people across the globe because they would be able to tamper with the data. At the same time, blockchain also supports data sharing.

1.13.3 Supply Chain Management

The desire to significantly improve the supply chain management can be achieved by blockchain technology. Recent advancements of the Internet of Things and blockchain technologies support much better supply-chain transactions. When the product is passed on from the manufacturer to the customer, the important data files are recorded in blockchain. Companies are now able to trace products and raw materials to improve their product quality.

1.13.4 Smart Transactions

Businesses are now able to establish smart contracts on blockchain, and these smart contracts are widely used to implement business collaborations in general and international business processes. Enterprises are now able to automate transactions based on smart contracts on block chains without manual confirmation.

1.14 Blockchain in Smart Cities

The rapid rate of increasing urbanization has given us considerable development in preparing long-term capable and productive solutions. Many cities have been plagued with management and maintenance of energy, waste, and several other government services. Blockchain can provide solutions to these problems. Technologies like machine learning, deep learning, Internet of Services, artificial intelligence, etc. are being utilized to convert cities into smart cities [66]. The main goal of such cities is to ease the living condition by utilizing the given resources to its best and minimizing the expense for the same. Blockchain can provide transparent and secure answers to the communication, financial, and healthcare sectors, which are key for the idea of smart cities.

1.14.1 Uses of Blockchain in Smart Cities

Blockchain for Cities [67], a United Nations initiative, is a program to bring together many technologies like BIM (Building Information Modeling), the GIS (Geographical Information Systems), and UPIS to inspire cities. This assists them to convert these cities into smart cities.

According to the researchers, some of the areas where smart cities utilize blockchain are as follows:

Security
Blockchain has proven to be a secure and safe medium for many tasks. The decentralized ledgers and hashing guards the data given by the users.

Energy
Using smart contracts has allowed renewable resources enormous development, along with financial support and energy policies [48]. For example, households with solar power can easily distribute any surplus energy with members of their grid.

Mobility
Since the privacy of the users is maintained, state departments can have an idea of the number of citizens who use travel in trains, cars, and buses and allocate the resources accordingly. The payment concerned with these transports can be done via blockchain-powered payment apps, to increase the overall efficiency of the system.

Waste
The waste produced can be tracked by blockchain technologies. Citizens who dispose of garbage correctly can be provided incentives and discounts to motivate other citizens. An efficient chain of collection, transportation, and disposal at landfills can be created, which will improve the recycling of waste, since the garbage at its origin will already be separated by the people themselves.

1.15 Issues and Constraints in Blockchain Application

This topic aims to grasp the challenges faced during the implementation of applications discussed above. We try to seek and rectify what problems and limitations are awaiting us in our future goals of solving healthcare-, architecture-, finance-, etc.,-related issues, on the basis of current prototypes and applications we have at our disposal.

Since most of the data in blockchain is immutable, backtracking and changing the data is a lot of work, hard forks are generally required for such transactions. The storing of these blockchain chains requires a massive amount of disk space. With this ever-growing size, users would not be able to download these chains, keeping in mind the load that would be required by networks to control the flow of data.

Since blockchain is a technology with ever-increasing efficiency, more research is required to open more doors for its applications in the sectors discussed. But before we work on the given application's real-world implementation, there are lots of legal and accountability issues to examine and review. The steep work put into project direction, data entry and monitoring, and billing and invoicing approaches has been found ancient and old-fashioned for the present age. Even with industry being conservative and intricate, it is time to modify the practices and habits. The acceptance of blockchain might prove to be an obstacle, but there is evidence that the advantages and gains connected are massive [68].

1.16 What the Future Holds for Blockchain

Blockchain is still in an early stage of development, so it still has mixed predictions about its future and potential.

A study conducted by the analytical firm Gartner concluded that:

- It is estimated that only 10% of the companies would have achieved complete transformation by the year 2022.
- Businesses would be worth $10 billion by the year 2023.
- It is estimated that by 2026, the business value added by blockchain would be around $360 billion, and by 2030, the amount will shoot up to $3.1 trillion.

Cybersecurity is the most promising area of growth for blockchain technology. The most significant challenge that businesses face is hacking or data tampering. By the use of blockchain technology, hacking can be prevented and it allows the participants to verify the file's originality.

The International Data Corporation (IDC) has reported that many IoT corporations have considered the execution of blockchain technology in their solutions.

This is due to the fact that blockchain technology provides the most secure foundation for communication between different IoT devices. Many existing protocols have failed when they have been applied to IoT devices; thus blockchain technology has proven to be effective.

The idea of the distributed ledger is very efficient for government agencies that have to govern huge amounts of data. At present all the agencies have separate databases so in order to gain information about residents, they have to constantly contact each other.

However, after the execution of blockchain technologies, the functioning of such agencies will improve and they will produce larger outputs.

Estonia has already started using blockchain technology on the governmental level. It is believed that all the public services in Estonia have access to X-Road, a dispersed digital ledger that has all the information about its residents. This technology is a very high-level encryption technology that includes two-factor authentication, and this enables the people to control their own data and be sure of its security.

In the future, blockchain will completely change the complexion of trade processes in many industries, but adopting/implementing it will require a lot of time. We can expect that governments around the world will finally accept the benefits of blockchain technology and will start using it for improving financial and public services. Though many blockchain startups will fail as well, people will get more experience and knowledge on how to implement this technology. Blockchain will encourage people to adopt new skills, whereas traditional businesses will have to completely rethink their policies [69].

References

1. Bitcoin: A peer-to-peer electronic cash system. https://bitcoin.org/bitcoin.pdf
2. A summary of research on blockchain in the field of intellectual property. https://www.sciencedirect.com/science/article/pii/S187705091930239X
3. A review paper on cryptography. https://www.researchgate.net/publication/334418542_A_Review_Paper_on_Cryptography
4. Amtenbrink, Fabian. (2011). Central Bank Challenges in the Global Economy. https://link.springer.com/chapter/10.1007/978-3-642-14432-5_2.
5. Del Río, César. (2017). Use of distributed ledger technology by central banks: A review. Enfoque UTE. 8. 1-13. 10.29019/enfoqueute.v8n5.175
6. Saucier, C., & McKibben, M. Ubitquity. Telephone interview. 30 June 2016.
7. Authentication using Smart Contracts in a Blockchain. https://publications.lib.chalmers.se/records/fulltext/256254/256254.pdf
8. Antonopoulos, A. M. Mastering bitcoin: Unlocking digital cryptocurrencies. https://static1.buchi.com/sites/default/files/webform/pdf-mastering-bitcoin-unlocking-digital-cryptocurrencies-andreas-m-antonopoulos-pdf-download-free-book-4a0a742.pdf
9. Avi Spielman. Blockchain: Digitally Rebuilding the Real Estate Industry. http://dspace.mit.edu/bitstream/handle/1721.1/106753/969450770-MIT.pdf?sequence=1
10. Kharif, O. Blockchain goes beyond crypto-currency. Bloomberg.com. Bloomberg, 19 May 2016. Web. 16 June 2016.
11. Nakamoto, S. Bitcoin: A peer-to-peer electronic cash system. Legitimate applications of peer-to-peer networks (n.d.): n. pag. Bitcoin.org. Web. 22 May 2016.
12. Lamport, L. (1985). A new solution for the byzantine generals problem. Decision support systems 1.2. p. 182. Microsoft. Microsoft. Web. 6 July 2016.
13. How does bitcoin work? FAQ. N.p., n.d. Web. 6 July 2016.
14. Proof of work. Bitcoin Wiki. N.p., n.d. Web. 6 July 2016.
15. Ethereum: State of knowledge and research perspective. https://fps2017.loria.fr/wp-content/uploads/2017/10/19.pdf
16. Central bank digital currency and the future of monetary policy. https://www.nber.org/papers/w23711
17. Automation processes and blockchain systems. https://papers.ssrn.com/sol3/papers.cfm?abstract_id=2890435
18. Blockchain solution for CBDC. https://consensys.net/blockchain-use-cases/payments-and-money/cbdc/
19. Blockchain in financial services. https://consensys.net/blockchain-use-cases/finance/
20. Central banks and future of digital money. https://pages.consensys.net/central-banks-and-the-future-of-digital-money
21. Applications of blockchain in banking and finance. https://www.researchgate.net/publication/327230927_Applications_of_Blockchain_Technology_in_Banking_Finance
22. Preferred reporting items for systematic reviews and meta-analyses: The PRISMA statement. https://journals.plos.org/plosmedicine/article?id=10.1371/journal.pmed.1000097
23. Blockchain technology in healthcare: A systematic review. https://www.mdpi.com/2227-9032/7/2/56/htm
24. Electronic medical records system. https://patents.google.com/patent/US5924074A/en
25. Personal health records: Definitions, benefits, and strategies for overcoming barriers to adoption. https://academic.oup.com/jamia/article/13/2/121/729326
26. Definition, structure, content, use and impacts of electronic health records: A review of the research literature. https://www.sciencedirect.com/science/article/abs/pii/S1386505607001682
27. What the GDPR means for businesses. https://www.sciencedirect.com/science/article/abs/pii/S1353485816300563

28. Supply chain management in health services: An overview. https://www.emerald.com/insight/content/doi/10.1108/13598541111127146/full/html?mobileUi=0&fullSc=1&mbSc=1&fullSc=1&fullSc=1&fullSc=1
29. Hitching healthcare to the chain: An introduction to blockchain technology in the healthcare sector. https://timreview.ca/article/1111
30. Transform blockchain into distributed parallel computing architecture for precision medicine. https://ieeexplore.ieee.org/abstract/document/8416392
31. Leveraging blockchain for retraining deep learning architecture in patient-specific arrhythmia classification. https://ieeexplore.ieee.org/abstract/document/8333451
32. Healthcare data breach statistics. https://www.hipaajournal.com/healthcare-data-breach-statistics/
33. Applications of how blockchain is reviving healthcare. https://builtin.com/blockchain/blockchain-healthcare-applications-companies
34. How blockchain will change construction. https://hbr.org/2019/07/how-blockchain-will-change-construction
35. Crossrail project: Finance, funding and value capture for London's Elizabeth line. https://www.icevirtuallibrary.com/doi/full/10.1680/jcien.17.00005
36. Blockchain technology in construction industry. https://www.designingbuildings.co.uk/wiki/Blockchain_technology_in_the_construction_industry
37. A systematic review of blockchain. https://jfin-swufe.springeropen.com/articles/10.1186/s40854-019-0147-z
38. Automated ledger account maintenance system. https://patents.google.com/patent/US5117356A/en
39. An empirical analysis of smart contracts: Platforms, applications, and design patterns. https://link.springer.com/chapter/10.1007/978-3-319-70278-0_31
40. Creation of smart-contracting collaborations for decentralized autonomous organizations. https://link.springer.com/chapter/10.1007/978-3-319-21915-8_1
41. Building automation system. https://patents.google.com/patent/US6967565B2/en
42. https://blockchainhub.net/dao-decentralized-autonomous-organization/
43. Data sharing and tracing scheme based on blockchain. https://ieeexplore.ieee.org/abstract/document/8593225
44. Byrne, P. Patrick Byrne, MIT Blockchain Independent Study Group. Group Meeting. 26 Apr 2016.
45. Madise, Ü., & Martens, T. E-voting in Estonia 2005. The first practice of country-wide binding Internet voting in the world. https://dl.gi.de/handle/20.500.12116/29155
46. Weiss, M., & Halyard, M. Voatz. https://www.hbs.edu/faculty/Pages/item.aspx?num=56024
47. Gupta, A., Kleinberg, J., Kumar, A., Rastogi, R., & Yener, B. Provisioning a virtual private network: A network design problem for multicommodity flow. https://dl.acm.org/doi/abs/10.1145/380752.380830
48. Andoni, M., Robu, V., Flynn, D., Abram, S., Geach, D., Jenkins, D., McCallum, P., & Peacock, A. Blockchain technology in the energy sector: A systematic review of challenges and opportunities. https://www.sciencedirect.com/science/article/pii/S1364032118307184
49. Kshetri, N., & Voas, J. Blockchain-enabled e-voting. https://libres.uncg.edu/ir/uncg/f/N_Kshetri_Blockchain_Enabled_2018.pdf
50. NirKshetri. Blockchain's roles in strengthening cybersecurity and protecting privacy. https://www.sciencedirect.com/science/article/abs/pii/S0308596117302483#!
51. Ayed, A. B. A conceptual secure blockchain-based electronic voting system. https://www.researchgate.net/publication/341498272_A_CONCEPTUAL_SECURE_BLOCKCHAIN-BASED_ELECTRONIC_VOTING_SYSTEM
52. Arcos, L. C. The blockchain technology on the music industry. https://www.researchgate.net/publication/327401244_The_blockchain_technology_on_the_music_industry
53. Atzei, N., Bartolett, M., & Cimoli, T. A survey of attacks on Ethereum smart contracts (SoK). https://link.springer.com/chapter/10.1007/978-3-662-54455-6_8

54. Ethnicity and political violence in Africa: The challenge to the Burundi state. https://www.sciencedirect.com/science/article/abs/pii/S0962629806000588
55. Wood, G. Ethereum: A secure decentralised generalised transaction ledger. https://files.gitter.im/ethereum/yellowpaper/VIyt/Paper.pdf
56. Open music initiative. https://en.wikipedia.org/wiki/Open_Music_Initiative.
57. Sitonio, C., & Nucciarelli, A. The impact of blockchain on the music industry. https://www.researchgate.net/publication/326225903_The_Impact_of_Blockchain_on_the_Music_Industry
58. Woloshyn, A. Imogen heap as musical cyborg: Renegotiations of power, gender and sound. http://alexawoloshyn.com/wp-content/uploads/2018/03/Alexa-Woloshyn-Imogen-Heap-copy.pdf
59. O'Dair, M. The networked record industry: How blockchain technology could transform the consumption and monetisation of recorded music. https://ualresearchonline.arts.ac.uk/id/eprint/14652/
60. Clohessy, T., Acton, T., & Rogers, N. Blockchain adoption: Technological, organisational and environmental considerations. https://link.springer.com/chapter/10.1007/978-3-319-98911-2_2
61. Hand, D. J., & Adams, N. M. Data mining. https://onlinelibrary.wiley.com/doi/abs/10.1002/9781118445112.stat06466.pub2
62. Amatriain, X. Big & personal: Data and models behind Netflix recommendations. https://dl.acm.org/doi/abs/10.1145/2501221.2501222
63. Mayfield, L. Voting fraud in early twentieth-century. Pittsburgh. https://www.jstor.org/stable/205101
64. Russom, P. Big data analytics. https://vivomente.com/wp-content/uploads/2016/04/big-data-analytics-white-paper.pdf
65. Hileman, G., & Rauchs, M. Global cryptocurrency benchmarking study. https://www.crowdfundinsider.com/wp-content/uploads/2017/04/Global-Cryptocurrency-Benchmarking-Study.pdf
66. Biswas, K., & Muthukkumarasamy, V. Securing smart cities using blockchain technology. https://www.researchgate.net/profile/Kamanashis_Biswas/publication/311716550_Securing_Smart_Cities_Using_Blockchain_Technology/links/5a2682d5a6fdcc8e866becfb/Securing-Smart-Cities-Using-Blockchain-Technology.pdf
67. Oliver, M., & Marsal, L. Blockchain4Cities. https://www.upf.edu/web/etic/blockchain4cities
68. Hashcash. Hashcash. N.p., n.d. Web. 7 July 2016.
69. Blockchain technology in the future: 7 predictions for 2020. https://www.aithority.com/guest-authors/blockchain-technology-in-the-future-7-predictions-for-2020/

Chapter 2
IoT-Based Unique Air and Noise Pollution Monitoring System

Pujan Soni, Kshitij Joshi, and Abhilasha Vyas

2.1 Introduction

In the entirety of this world, a majority of countries have agreed that global warming and climate change are the epitome of threats that this growing and expansive world faces [1]. But to that end, when these systems fail the citizens at the lowest form of cities and villages, the people are forced to take up situations into their own hands where they would test out and implement favorable and feasible solutions at their surroundings whether it is at colonies or private homes [2, 3]. The only thing they lack in tackling the problems of pollution is the fact that they do not have access to data that is captured and analyzed in real time. Hence in this chapter, we have designed a product which monitors the concentrations of air and sound pollution using IoT [4]. With specific reference to air pollution, the permissible limits are usually set by local and national governments, and although there are not direct limits for the general public, it is for separate types of industries and company which have to follow the guidelines, and when it comes to localities, the only limits are on the existence of emission certificates for vehicles. But even worse are the regulations on sound pollution which technically do not exist because there are not any, because there are only rules that after a certain time loud music and sounds are to be avoided which is believed to be ineffective due the infeasibility of enforcing such rules. That is where this product bridges the gap between infeasible solutions and active collection and analysis of data. Using IoT technology, the real-time monitoring of air and sound pollution concentration levels can be made. Real-time graphs and multiple LED bulbs indicating the intensity have been done using sensors (MQ-135 and LM-393), NodeMCU, Google Firebase, and the MIT Android Application. In this way, we have developed an effective method to keep track of air and sound pollution

P. Soni · K. Joshi (✉) · A. Vyas
GSFC University, Vadodara, Gujarat, India
e-mail: abhilasha.vyas@gsfcuniversity.ac.in

M. Moh et al. (eds.), *Smart IoT for Research and Industry*, EAI/Springer Innovations in Communication and Computing,
https://doi.org/10.1007/978-3-030-71485-7_2

Table 2.1 Guidelines issued from WHO for various categories [8]

Sr. no	Category	Year	Rule
1.	Household fuel combustion	2014	Clean fuels and technologies for household cooking, heating, and lighting
2.	Selected pollutants	2010	Lessening well-being dangers from presentation to encompassing outflows of gases and synthetic compounds that may penetrate and gather inside, just as from synthetic substances that might be utilized in building materials or goods that add to indoor air contamination
3.	Dampness and mold	2009	Well-being impacts from indoor dampness and introduction to organisms, for example, form, growths, and microbes which produce spores into indoor air
4.	Particulate matter, ozone, nitrogen dioxide, and sulfur dioxide	2005	Decreasing the impacts on soundness of air contamination

levels. The execution and improvement of the observing framework affirms the capacity of such a framework to permit the normal open approach to checking and brings about continuous methods for the Internet of Things gadgets. The various segments, for example, microcontrollers, sensors, programming, and remote modules, that are utilized to execute the air quality and commotion contamination checking framework are portrayed in this section. This framework shows the ongoing checking information and graphical portrayal of the contamination alongside air quality file viewpoints. Continuous readings, obtained from the microcontroller, are legitimately passed on Google Firebase Console with the goal that it is accessible to any client over the world that is associated with the web [5–7].

WHO sets suggested limits for health hurtful convergences of key air toxins both outside and inside structures and homes, in view of worldwide union of logical proof. WHO rules spread yearly and everyday groupings of fine particulates, nitrogen dioxide, sulfur dioxide, carbon monoxide, and ozone. WHO issued rules additionally spread clamminess in year 2009 and outflows of gases and synthetic substances from decorations and building materials that gather inside in year 2010. Most as of late, WHO rules for indoor air quality, family unit fuel burning, set cutoff points on emanations from cooking and warming ovens, just as suggestions with respect to clean fuel use (Table 2.1).

2.2 Internet of Things (IoT)

The world is discussing the Internet of Things connected to the Internet. Tragically, the conversation remains covered in a fog of vulnerability for some leaders. Like other progressive innovation headways, IoT creates a lot of merited buzz, making it extreme to get straight [9, 10].

The Internet of Things can incorporate everything from an assembling control framework that runs a sequential construction system to the ecological sensors that keep transient stock at unequivocally the correct conditions inside cargo terminals. [11]. While workers organize switches and other customary registering gadgets assume indispensable jobs in the advanced web of things, for example, directing traffic, adjusting utilization, and leading help quality observing, they are not the headliner [12]. The genuine activity lies with the equipment that produces, dissects, and stores the information.

2.2.1 Advantages of IoT Over Other Technologies

The Internet of Things can join everything from a gathering control structure that runs a successive development framework to the natural sensors that keep transient stock at unequivocally the right conditions inside load terminals. While laborers sort out switches and other standard enrolling contraptions accept key positions in the advanced internet of things, for instance, coordinating traffic, changing usage, and driving assitance quality noticing, they are not the main event. The genuine movement lies with the hardware that produces, analyzes, and stores the data [13–15].

Surrounding knowledge and self-sufficient control to coordinate the ideas of the IoT and independent control, with starting results toward this course considering objects as the main force for self-governing IoT. Preparing an operator (i.e., IoT gadget) to carry on cleverly a fortification learning approach, a learning specialist can detect the earth's state (e.g., detecting home temperature), perform activities (e.g., turn air-conditioning on or off), and learn through the boosting amassed rewards it gets in the long term [16].

2.2.2 Need of Air and Noise Quality Monitoring System

In today's era, growing issues include global rising temperature, ozone layer depletion, water resources contamination, and factors like air pollution and noise pollution, etc. Among all these major pollutions, air pollution and noise pollution play a major role in making our lives vulnerable. In order to the face worst case scenarios, monitoring and measuring plays a vital role to eradicate their effects. Air contamination in significant urban communities across India arrived at disturbing levels in the past few years, especially in Delhi where inhabitants face a general well-being crisis. The national capital locale has been covered in a thick cover of exhaust cloud, constraining a great many people to remain inside. Flight activities at the Delhi air terminal were likewise influenced because of the low perceivability brought about by the air contamination. In mid of 2019 the air quality file in Delhi dropped to 999 throughout the end of the week, making it the most contaminated capital city on the planet. While as far as possible set by the World Wellbeing Association as 25, anything over 500 represents a genuine hazard to the respiratory frameworks of everybody. Air Quality Index or AQI measures the concentration of PM 2.5 levels – fine particles of less than 2.5 microns that can enter the bloodstream and penetrate the lungs and heart, which are linked to chronic respiratory diseases. Not exclusively does the harmful air influence sound lungs, it is additionally found to cause unexpected losses among those experiencing respiratory diseases. As indicated by Greenpeace, more than 1.2 million Indians kicked the bucket rashly in 2017 because of air contamination. One of the significant purposes behind Delhi's poisonous air is accepted to be the consumption of homestead squander in neighboring Punjab and Haryana. Delhi government authorities have considered the city a "gas chamber" and accused the state legislatures of Punjab and Haryana of not doing what is needed to control the illicit homestead fires. A critical drop in AQI levels was seen post-Diwali where firecrackers assumed a mind-boggling job in making a poisonous mixed drink. Different urban areas where contamination levels have coordinated up to Delhi's are Lucknow, Patna, and Kolkata. While numerous online networking clients compared the air contamination in India to smoking cigarettes, a few clients shared screen captures of a constant air quality list which indicated the contamination level in the number cigarettes smoked [17–21].

On summing up the current contextual analysis, we can watch the issues of rise in pollutants present in air quality level resulting in decrement of air quality level and augmentation in sound level which requires deep research. With the advancement of IoT, a solution for such problems can be designed. As per the unpredictable weather conditions in today's scenario, there is a need for a system which can keep track of changing states of certain weather parameters. This research work is based on implementation of IoT for monitoring important parameters like air and noise levels with alert systems [22, 23].

2.3 Literature Review

In [24] the system is a sound pollution sensor used for measuring noise in the range of 30–130 dB. Data can be accessed only on a handheld module (offline) and it consumes moderate power. Benefits include being able to change the measuring rate from slow to fast.

The item in [25] is a class II sound level, sound weight level (SPL), decibel (dB), or commotion meter with perceptible or visual caution yield ability. Information can be gotten just on a divider-mounted module (disconnected). This item is utilized for constant checking of clamor in modern conditions, for example, foundries, industrial facilities and assembling creation floors.

Reference [26] is an indoor air quality monitor designed for small cooling and heating air handling systems in industrial, residential, office, and commercial facilities. The IAQ monitor has an LCD display, showing the temperature, humidity, and carbon dioxide values in succession of the area under surveillance. Advantages include comparatively low power consumption and moderately difficult to calibrate.

In [27] the item quantifies VOC (unpredictable natural mixes) and formaldehyde continuously. Its advantages include the following: no online information openness and portable alarms given, quick reaction time, minimized, low force utilization, simple to convey, and superb for checking the indoor air nature of homes, workplaces, production lines, inns, schools, and other indoor situations.

2.3.1 Existing Technologies for Sound Monitoring System

Sound level monitoring systems use decibel units for measuring the intensity of sound. Decibel values are gotten by utilizing an instrument called the sound-level meter. There are different sorts of sound-level meters accessible, some with more exactness and adaptability. Be that as it may, all meters comprise four fundamental parts: microphone, input amplifier, output amplifier, and a readout device.

2.3.2 Existing Technologies for Air Monitoring System

Observing and estimating air quality is done by a wide range of associations, for instance, the administration, nearby boards, industry, research bodies, and ecological pressure groups. Air quality is measured with the Air Quality Index, or AQI. The AQI works like a thermometer that runs from 0° to 500°. In any case, rather than demonstrating changes in the temperature, the AQI is a method of indicating changes in the measure of pollution in the air noticeable all around. Air quality is a proportion of how spotless or dirtied the air is. An alternate sort of instrument is expected to quantify every contamination. Every significant town and huge urban community has at any rate one air checking station that contains exceptionally particular and fragile instruments for estimating a scope of contaminations. These destinations are about the size of a little band and are pricey to purchase and maintain. A broadly utilized strategy for estimating one of the principle air toxins from street traffic, nitrogen dioxide, utilizes a basic gadget called a dispersion tube. Most air contamination estimating gears are called dynamic gadgets since they utilize a siphon to suck air into them [28–34].

2.3.3 Research Gaps

These are the accompanying examination holes which we have distinguished after efficient writing audit:

1. The current frameworks were inadequate in the field of reasonableness and effectiveness of activity.

2. It was discovered that the cost factor in execution of these frameworks was altogether higher than the continuous development innovation-based frameworks.
3. Putting away the information in these frameworks was very troublesome undertaking when contrasted with current frameworks.

2.3.4 *Objective of the Research*

This research aims to build a unique air and noise pollution monitoring system (AIR-O-SPY) based on IoT to help large- and small-scale users to ensure optimum ambience with the suggestion of feasible and cost-effective solutions. The monitoring system would gather data from multiple sensors and correlate data with a wide range database that communicates directly to the user via the mobile android application [35].

2.4 Proposed Model

This section presents a proposed model based on IoT. This model is using hardware for collection of data and softwares for analyzing data. The proposed model is divided into two phases. First phase is the data collection phase in which field devices like sensors are employed at the field, and in the second phase collected information will be analyzed as per proposed algorithm. This model is designed to identify the air quality level and sound level in environmental. Flowchart 1 shows the data processing and data collection phases of the proposed model (Fig. 2.1).

Proposed Algorithm (Fig. 2.2)

2.5 Implementation

2.5.1 *Hardware*

The hardware setup of the proposed model consists of equipment like sensors, microcontroller, and jumper connectors. For this study the air quality sensor MQ-135 and the sound sensor LM-393 are used. Wi-Fi-enabled microcontroller is used to store collected data on cloud. The upcoming subsections are describing the hardware used for implementation of this model [36].

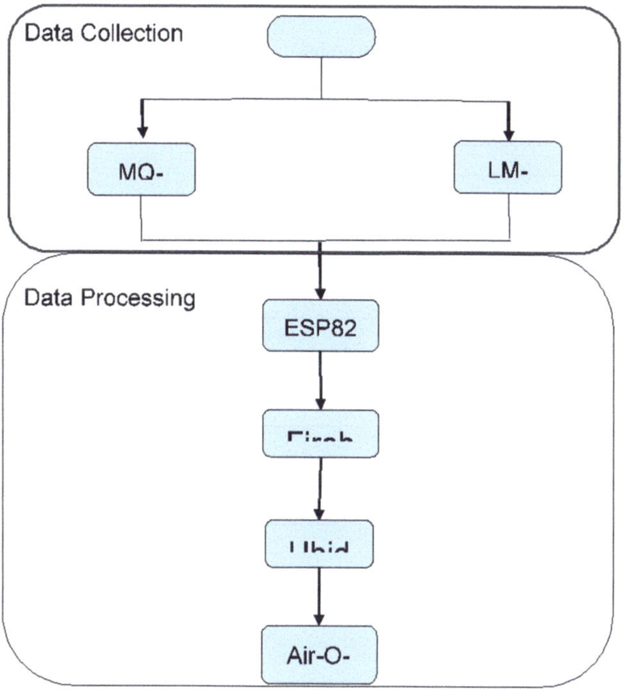

Fig. 2.1 Data collection phase and data processing phase of the proposed model flowchart

2.5.1.1 Sensors

An electronic gadget which quantifies a physical property and records, demonstrates, or reacts to its change is called a sensor. Based on the parameters being measured, sensors can be categorized as temperature sensor, proximity sensor, light sensor, smoke sensor, etc. The sensors being used in this product are MQ-135 and LM-393 for air and sound pollution.

MQ-135

MQ-135 gas sensor has high affectability to alkali, liquor, CO2, CO, and smoke. The yield voltage changes as indicated by the convergence of the deliberate gases. It utilizes 5.0V ± 0.2V DC. It is known for its quick reaction and recuperation, customizable affectability, long life, and minimal effort. Applications incorporate air pollution identifier, liquor breath analyzer, air quality screen, and for substance investigation (Table 2.2) [37–39].

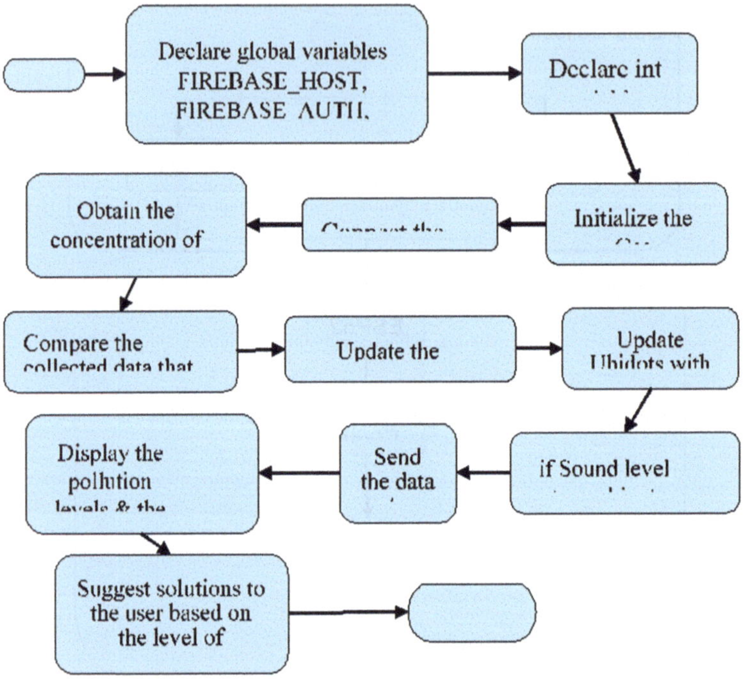

Fig. 2.2 Flowchart of model

Table 2.2 Pin configuration for MQ-135 gas sensor module [40]

Pin no	Pin name	Description
1.	Vcc	Used to power the sensor, generally the operating voltage is +5V
2.	Ground	Used to connect the module to system ground
3.	Digital out	You can also use this sensor to get digital output from this pin, by setting a threshold value using the potentiometer
4.	Analog out	This pin outputs 0–5V analog voltage based on the intensity of the gas

LM-393

LM-393 sound detection sensor is utilized to recognize the force of the sound on the earth. It works on 3.3–5V DC. LM-393 sound detection sensor module for Arduino identifies whether sound has surpassed a limit esteem. The sound recognized by a microphone is fed into a LM-393 op-amp. The sound level set point is balanced by means of an onboard potentiometer. At the point when the sound level surpasses the set point, an LED on the module is lit up and the yield is sent LOW. Yield is computerized and simple to introduce. The primary application is as a sound locator wherein clamors in various decibel levels can be estimated (Fig. 2.3, Table 2.3) [41, 42].

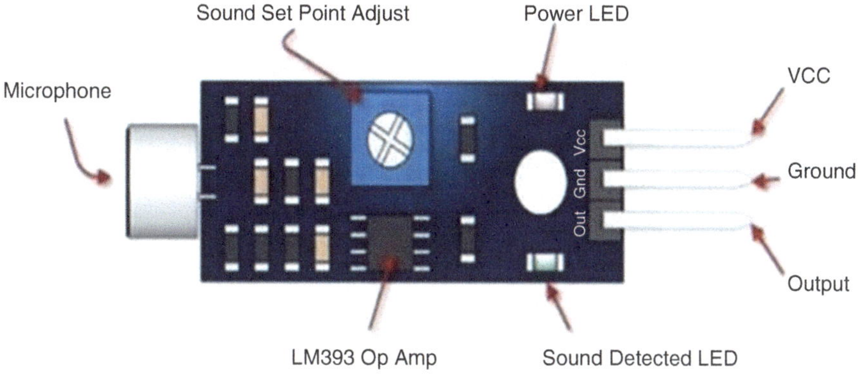

Fig. 2.3 Pin configuration for LM-393 sound sensor module [43]

Table 2.3 Pin configuration for *LM-393* sound sensor module [44]

Pin number	Pin name	Description
1	OUTPUT1	Output of Op-Amp 1
2	INPUT1−	Inverting input of Op-Amp 1
3	INPUT1+	Non-inverting input of Op-Amp 1
4	VEE, GND	Ground or negative supply voltage
5	INPUT2+	Non-inverting input of Op-Amp 2
6	INPUT2−	Inverting input of Op-Amp 2
7	OUTPUT2	Output of Op-Amp 2
8	VCC	Positive supply voltage

2.5.1.2 Microcontroller

It is a cost-efficient microcontroller with full TCP/IP stack and accessibility to Wi-Fi. In our project, we use NodeMCU which has hardware based on ESP-12 module. ESP8266 core for Arduino IDE has gotten one of the main software development platforms for the various ESP8266 modules and for the advancement board, industry NodeMCU [45, 46].

Due to unavailability of real-time air and sound pollution monitoring systems and cost-effective and applicable solutions to tackle this growing problem, it is very difficult to survive for long with pristine health and ambient conditions that are safe and secure for the prosperity of future generations. In order to reduce the hassle that common citizens have to take to resolve such issues like attempting to consult with authorities and waiting for long periods of time for solutions to be implemented and to achieve the same, we use multiple sensors, namely, MQ-135 air quality sensor and LM-393 sound sensor, to be controlled by the ESP8266 NodeMCU working in tandem with Firebase which would act as the active database collection terminal and Ubidots with an active graph on display with an Android mobile application developed in the MIT App Inventor 2. This application would show the live levels of the quality of air and sound. And once the set thresholds are passed, an alert would pop up along with a solution that can be implemented to reduce the effects of the situation that has arisen.

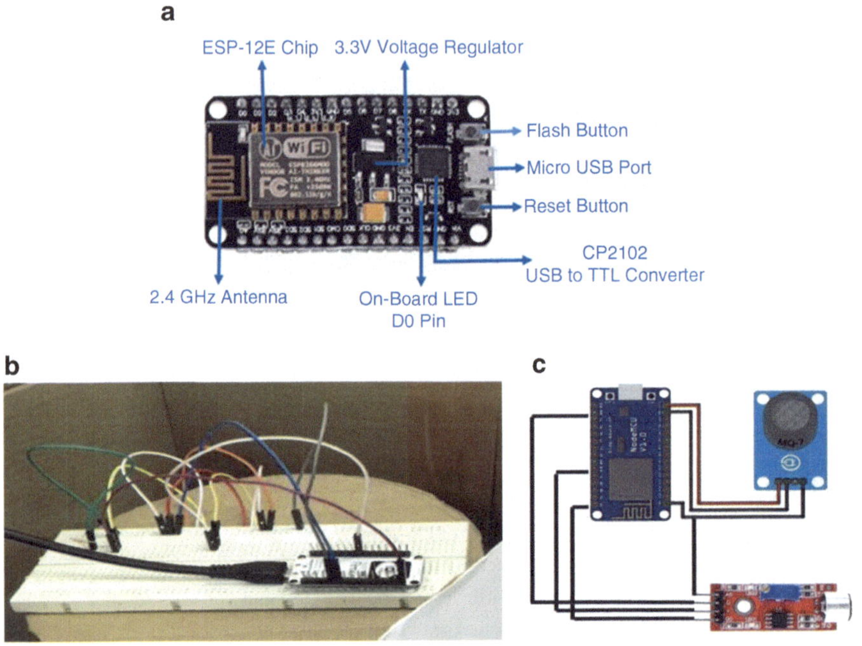

Fig. 2.4 (**a**) Pinout of NodeMCU. (**b**) Hardware setup. (**c**) Circuit diagram [47]

Figure 2.4b shows the setup for a unique air and noise pollution monitoring system (AIR-O-SPY) based on IoT. The MQ-135 and LM-393 sensors are used to measure air quality and sound levels are connected to the ESP8266 microcontroller. The two sensors are placed in the simulation environment to detect the changes in air and sound ambience present in the environment (AIR-O-SPY). This simulated environment is set up to create a real-time dataset for detection of air and noise pollution level. The real-time data is displayed on the screen of the AIR-O- SPY android application and on Firebase with an active graph displayed by Ubidots. We are able to see applicable solutions depending on the levels of sound and air pollution.

The following figure is the circuit connections for a unique air and noise pollution system that continually monitors the parameters of the field.

2.5.2 *Software*

2.5.2.1 Firebase

Firebase Cloud Messaging is a cross-stage messaging arrangement that permits you to convey messages at no expense. You can send information to the microcontroller using the Firebase and vice versa. The data obtained on Firebase can be displayed on the app along with warning messages. The users can then take action from a remote place and maintain the quality of the plants. It acts as a link between the app and the NodeMCU. Ubidots is used to relay live graphs that are trackable when it comes to air quality (Fig. 2.5) [48–50].

2.5.2.2 Ubidots

Ubidots is an Internet of Things (IoT) data analytics and visualization company. Various IoT-based applications can be built with the help of interactive, real-time visualization or also known as widgets. It can convert the analytical data into a graphical form which can be understood easily by the user (Fig. 2.6) [52, 53].

Fig. 2.5 Real-time database: Google Firebase Console [51]

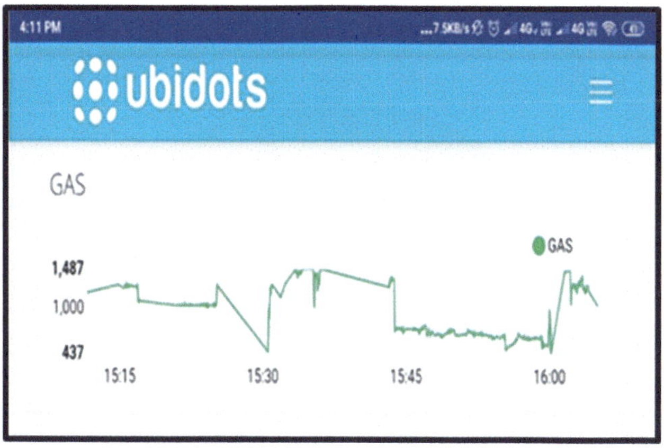

Fig. 2.6 Line graph represents the real-time data of air quality in Ubidots [54]

2.5.2.3 Android App

For the data to be accessed using the mobile app, data is pushed from the sensor to the ESP8266 microcontroller which in turns sends it to the Firebase database on the cloud. This data can be accessed using the android apps built on [55] MIT app inventor having the permission to access this database. These android apps can be installed on the devices of the operator so that he/she can access the data from the operating unit being anywhere in the world (Fig. 2.7).

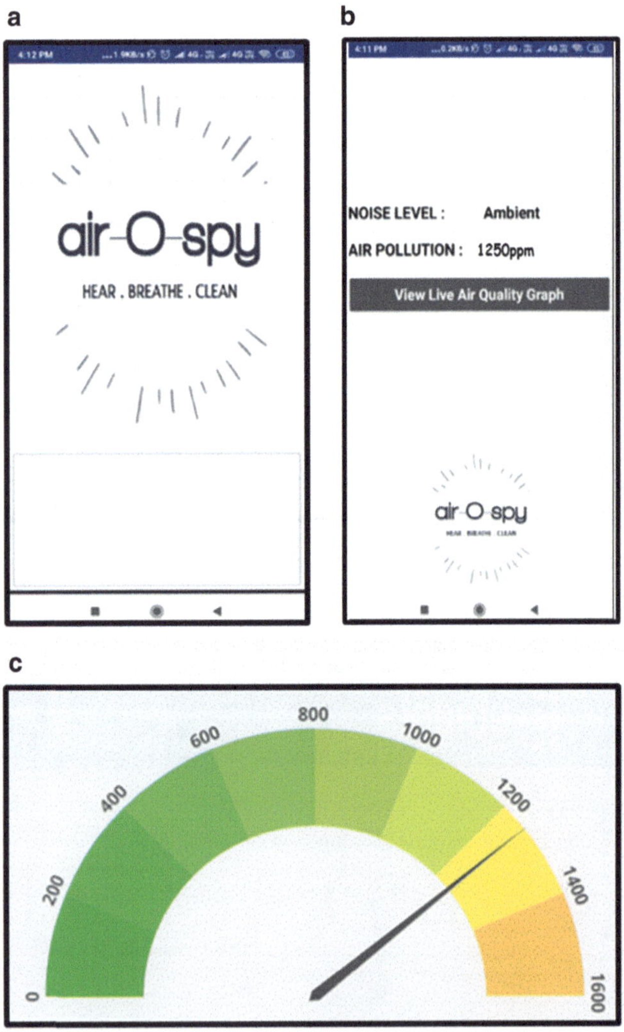

Fig. 2.7 (**a**) Page one of android app's UI. (**b**) Page two of android app's UI. (**c**) Page three of android app's UI showing representation of the data in gauge meter [56]

2.6 Experiments and Results

Using our smart air and sound pollution monitoring system (AIR-O-SPY), we are able to achieve real-time data collection of air and sound pollution levels of the simulated environment. The real-time data is displayed on the screen of the AIR-O-SPY android application and on Firebase with an active graph displayed by Ubidots. We are able to see applicable solutions depending on the levels of sound and air pollution.

Tables 2.4 and 2.5 describe the threshold level specified by the certain standard authorized organizations. We are using these threshold values to compare with the results to validate and check the efficiency of the proposed model. The system obtains the values and compares them with the given reference tables for redirecting them to separate levels. Table 2.6 shows the few results of our model tested in different locations of Vadodara city, Gujarat.

2.6.1 *Research Contribution*

Special air and noise pollution monitoring system using IoT innovations will assist basic residents with detecting and diminishing air and noise pollution and improving efficiency. The results of the proposed approach will help in increasing the life

Table 2.4 Different levels of air quality according to different value ranges of Air Quality Index (AQI) given by standard authority [57]

Air quality index levels of health concern	Numerical value (AQI)	Meaning
Good	0–50	Air quality is viewed as acceptable, and air contamination presents next to zero hazard.
Moderate	51–100	Air quality is satisfactory; in any case, for certain pollutants, there might be a moderate health concern for few individuals who are surprisingly touchy to air pollution
Unhealthy for sensitive groups	101–150	Individuals from touchy gatherings may encounter wellbeing impacts. The overall population is not probably going to be influenced
Unhealthy	151–200	Everyone may begin to experience health issues
Very unhealthy	201–300	Health alert; everybody may encounter more serious health impacts
Hazardous	301–500	Health admonitions of emergency conditions. The whole populace is bound to be affected

Table 2.5 Threshold levels for sound at different types of areas during day and night again given by the standard authority

Area	Daytime (dB)	Nighttime (dB)
Industrial areas	75	70
Commercial areas	65	55
Residential areas	55	45

Table 2.6 Result table with different real data representations

Sr. no	Location	Threshold value (AQI)	Obtained value (AQI)	Level
1.	GSFC University (city outskirts)	0–50	62	Moderate
2.	GSFC Limited (city outskirts)	101–150	137	Unhealthy for sensitive groups
3.	Vadodara International Airport	101–150	119	Unhealthy for sensitive groups
4.	Kamatibaug Garden Vadodara	0–50	47	Good
5.	Makarpura GIDC (city outskirts)	150–250	204	Unhealthy

Note: The average AQI of the Vadodara City is recorded as 64 AQI (15th August 2020) via [58]

expectancy of the individuals living or working in such weak conditions. To eradicate the flaws of the existing technologies, this system has come up with unique solutions which are as per the following:

Free limitation on vitality utilization. The sensor hubs are normally fueled by batteries with huge limits or vitality collect gadgets or electrical cable.

No finding gadget. The area of a sensor hub is known once it was sent since the sensor hub is stationary.

Loose confinements on weight and size. The transporter of the sensor hub can convey adequate enough weight.

Multiple sensors per hub. One sensor hub can outfit with a few kinds of sensors in view of the free restrictions on weight and size.

Exact and dependable information. Sensor hub can incorporate with helping devices on account of the free constraints on weight and size.

Ensured arrange availability. When the fixed sensor hub joined the system, the geography is fixed and the network is ensured.

All around adjusted and looked after sensors. The sensor hubs can be very much aligned and kept up by the experts intermittently [59].

2.7 Conclusion and Future Scope

We have successfully implemented the AIR-O-SPY through IoT. Using the real-time data obtained real time data collection of air and noise pollution levels of the simulated environment. The citizens are sent out reliable and feasible solutions depending on the level of threat. In the future this system can be directly integrated into further technology like air controlling and noise canceling systems [15, 60].

With appropriate precision the information can likewise be utilized by government associations to watch out for assembling based ventures climate; they release ideal measure of pollutants noticeable all around or not, and they can likewise utilize the sound checking system to watch the sound created by the manufacturing units.

In the future such systems can be implemented to monitor the equivalent for the health concerns of the employees and laborers. Apart from the industrial sector, this system can likewise be utilized by household individuals who are cautious in regard to their health and safety to monitor air and sound level in their homes or residential areas.

Acknowledgment We would like to acknowledge the Department of Science and Technology, Gujarat Council of Science and Technology, Gujarat Government for setting up a design lab equipped with IoT devices in GSFC University, Vadodara, with the help of which we were able to complete this study.

References

1. https://www.coolingindia.in/air-quality-hvacr-industry-articles-air-pollution-quality-index-effects-reduce-air-pollution-heating-ventilation-air-conditioning-refrigeration-ac-ventilation-refrigeration-air-quality-chillers-cold-s/indoor-air-quality-hvacr-industry-air-quality-testing-air-quality-monitor-leed-certification-air-test-green-building-concept-air-purifier-indoor-air-pollution-clean-air-quality-pollution-breathe-roof/page/3/
2. https://www.wikihow.com/Test-the-Air-Quality-in-Your-Home
3. https://cleanup.expert/info/how-to-test-indoor-air-quality/
4. https://www.iqsdirectory.com/resources/benefits-of-air-pollution-control/
5. https://circuitdigest.com/microcontroller-projects/iot-air-pollution-monitoring-using-arduino
6. https://www.iotchallengekeysight.com/2019/entries/smart-land/211-0515-025039-real-time-air-quality-monitoring-system-based-on-iot
7. https://www.skyfilabs.com/project-ideas/air-quality-monitoring-system
8. https://www.who.int/airpollution/guidelines/en/
9. Agrawal, R., Paprzycki, M., & Gupta, N. (Eds.). (2020). *Big data, IoT, and machine learning: Tools and applications*. Boca Raton: CRC Press.
10. https://www.ijitee.org/wp-content/uploads/papers/v8i9S2/I11160789S219.pdf
11. https://davra.com/what-is-iot-internet-of-things/
12. https://www.sureworks.in/index.php/services/digital-services/internet-of-things
13. Bhushan, D., & Agrawal, R. (2020). The internet of things: Looking beyond the hype. In *An industrial IoT approach for pharmaceutical industry growth* (Vol. 2, p. 231).
14. Bhushan, D., & Agrawal, R. (2020). Security challenges for designing wearable and IoT solutions. In *A handbook of internet of things in biomedical and cyber physical system* (pp. 109–138). Cham: Springer.
15. Kumar, A., Srinivas Kumar, P., & Agarwal, R. (2019). A face recognition method in the IoT for security appliances in smart homes, offices and cities. In *2019 3rd international conference on computing methodologies and communication (ICCMC)*. IEEE.
16. https://www.researchgate.net/publication/336186012_IoT-Based_Real_Time_Air_Pollution_Monitoring_System
17. https://www.sciencedirect.com/topics/earth-and-planetary-sciences/air-quality-monitoring
18. https://oxfordshire.air-quality.info/why-air-quality-is-important#:~:text=It%20provides%20air%20quality%20standards,the%20air%20that%20they%20breathe
19. https://www.eea.europa.eu/articles/cleaner-air-benefits-human-health
20. http://www.ijirset.com/upload/2018/may/136_17_IOT.pdf

21. https://www.ppsthane.com/blog/benefits-of-air-quality-monitoring#:~:text=Area%20having%20low%20air%20exchange,less%20sources%20of%20air%20pollution
22. https://www.hackster.io/east-west-university/indoor-air-quality-monitoring-system-5b5244
23. https://www.researchgate.net/publication/328015436_A_Smart_Air_Pollution_Monitoring_System
24. Testo 816-1 sound level meter, "Testo SE and Co. KGaA". Website: www.testo.de
25. SLT-TRM sound detection sensor, PCE Instruments TM Registration at PONS (Alicante, Spain). https://www.pce-instruments.com/english/measuring-instruments/test-meters/sound-level-meter-noise-level-meter-pce-instruments-sound-level-meter-pce-slt-trm-det_56247.htm
26. AI-IAQ by Ace instruments. https://acefirst.com/indoor-air-quality-moniotr-ai-iaq.php#:~:text=ACE%20Indoor%20Air%20Quality%20monitor,of%20the%20area%20under%20surveillance.
27. Extech VFM200 VOC detector by Extech instruments. Website: http://www.extech.com/products/resources/VFM200_UM-en.pdf
28. https://www.ncbi.nlm.nih.gov/pmc/articles/PMC4721779/#:~:text=Disadvantages%20or%20Challenges%3A&text=Inconveniences%20of%20calibration%20and%20maintenance,of%20urban%20surface%20is%20monitored
29. https://www.envirotech-online.com/article/air-monitoring/6/environmental-instruments/the-challenges-and-benefits-of-local-air-quality-monitoring/2104
30. https://www.intechopen.com/books/wireless-sensor-networks-insights-and-innovations/low-cost-energy-efficient-air-quality-monitoring-system-using-wireless-sensor-network
31. https://www.researchgate.net/publication/334092485_An_IoT_Based_Air_Pollution_Monitoring_System_for_Smart_Cities
32. https://www.academia.edu/35085518/IOT_BASED_AIR_POLLUTION_MONITORING_SYSTEM_USING_ARDUINO
33. http://www.ir.juit.ac.in:8080/jspui/bitstream/123456789/15827/1/SP13395_Pankhuri%20Arora_Nayanabh%20Chakravarty_IT_2018.pdf
34. https://nevonprojects.com/home-air-quality-monitoring-system-project/
35. Chen, S., Xu, H., Liu, D., Hu, B., & Wang, H. (2014). A vision of IoT: Applications, challenges, and opportunities with China perspective. *IEEE Internet of Things Journal, 1*(4), 349–359.
36. https://nevonprojects.com/iot-air-sound-pollution-monitoring-system/
37. https://www.olimex.com/Products/Components/Sensors/Gas/SNS-MQ135/resources/SNS-MQ135.pdf
38. https://datasheetspdf.com/pdf/605076/Hanwei/MQ-135/1
39. https://www.waveshare.com/w/upload/2/24/MQ-135-Gas-Sensor-UserManual.pdf
40. https://components101.com/sensors/mq135-gas-sensor-for-air-quality
41. https://www.alldatasheet.com/view.jsp?Searchword=Lm393%20datasheet&gclid=EAIaIQobChMIg7ic7duS6wIVC66WCh3GmwQEEAAYASAAEgLAx_D_BwE
42. https://www.electroschematics.com/lm393-lm2903-datasheet/
43. https://robu.in/product/lm393-sound-detection-sensor-module-black/
44. https://components101.com/ics/lm393-low-offset-voltage-dual-comparators
45. https://www.espressif.com/en/products/socs/esp8266
46. https://components101.com/wireless/esp8266-pinout-configuration-features-datasheet
47. https://components101.com/development-boards/nodemcu-esp8266-pinout-features-and-datasheet
48. https://firebase.google.com/?gclid=EAIaIQobChMIxYP8ltyS6wIVl7aWCh2K6A8lEAAYASAAEgJdQ_D_BwE
49. https://medium.com/firebase-developers/what-is-firebase-the-complete-story-abridged-bcc730c5f2c0
50. https://www.tutorialspoint.com/firebase/index.htm
51. https://console.firebase.google.com/u/0/
52. https://www.pubnub.com/integrations/ubidots/

53. https://www.iotone.com/software/ubidots-iot-application-development-platform/s71
54. https://ubidots.com/
55. https://appinventor.mit.edu/
56. https://iot-fpms.fandom.com/wiki/Ubidots
57. https://scijinks.gov/air-quality/#:~:text=Air%20quality%20is%20measured%20with,of%20pollution%20in%20the%20air.&text=Air%20quality%20is%20a%20measure,or%20polluted%20the%20air%20is
58. https://www.iqair.com/us/india/gujarat/vadodara
59. Zhang, L. Four challenges to be considered when developing IoT devices. https://dzone.com/articles/four-challenges-to-be-considered-when-developing-i
60. Zanella, A., Bui, N., Castellani, A., Vangelista, L., & Zorzi, M. (2014). Internet of things for smart cities. *IEEE Internet of Things Journal, 1*(1), 22–32.

Chapter 3
New Frontiers in Managing and Controlling Industrial Processes Through IoT

Neeraj Kumar Pandey and Ajitesh Kumar

3.1 Introduction

The Internet of Things (IoT) represents the environment of the different heterogeneous interconnected devices in the field of health, business, home utilities, security, logistic, manufacturing and computing and various real-time utilities with ubiquitous computing resources [1]. This set requires the network linkage to interconnect all IoT devices and a centralized resource provider to work as a backbone of this infrastructure. The Cloud-centric Internet of Things (CIoT) facilitates this service model with cloud computing services and IoT environment. In this architecture of cloud, the physical resources and devices are represented in the form of web resources. In the interconnection of physical devices, various non-computing devices are used like actuators, sensors and readers (wired and wireless mode) to communicate with each other while the rest of the services are taken from the cloud. All analytics is taken place on the cloud which makes the entire IoT network more decisive. As per the Moore's law, the size of the devices will get smaller and the circuit will get denser in the future. The major challenges in the CIoT architecture are BLURS [2], i.e. bandwidth, latency, uninterrupted, resource-constraint and security. These challenges need to be optimized as the number of connected devices is growing exponentially in the modern computing environment. The IoT environment is enabled with different computing environments like cloud computing [3], fog computing and edge computing. These computing environments provide backbone support to the IoT infrastructure and help in managing different services as depicted in Fig. 3.1.

N. K. Pandey (✉)
School of Computing, DIT University, Dehradun, Uttarakhand, India

A. Kumar
Department of CEA, GLA University, Mathura, Uttar Pradesh, India
e-mail: ajitesh.kumar@gla.ac.in

M. Moh et al. (eds.), *Smart IoT for Research and Industry*, EAI/Springer Innovations in Communication and Computing,
https://doi.org/10.1007/978-3-030-71485-7_3

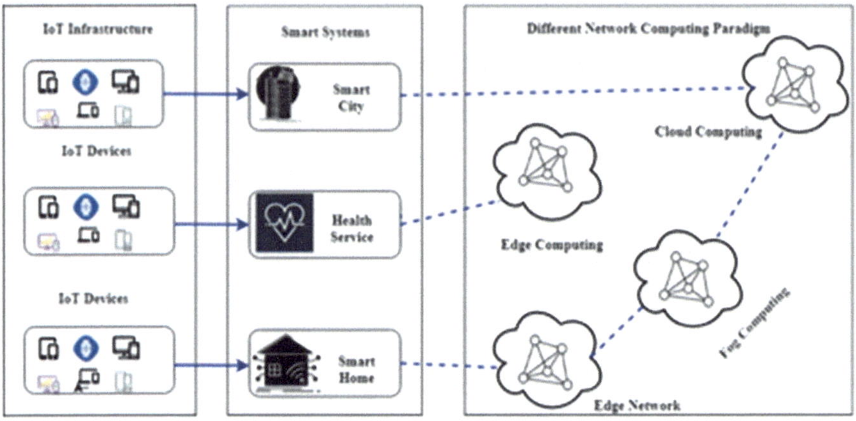

Fig 3.1 IoT computing environments with supporting computing paradigms

3.2 Background

The word Internet of Things (IoT) was first mentioned by Kevin Ashton, the Executive Director of Auto-ID Labs at MIT, who was the first to depict the Internet of Things while making an introduction for Procter and Gamble.

Before Kevin Ashton, Neil Gershenfeld wrote a book *When Things Start to think* [4]. In his book he gave a clear vision that IoT will head in the future. Kevin Ashton used Radio Frequency ID (RFID) for the Internet of Things. He thought if every device would have a unique identity, then it will be easier to stock them. Nowadays we use RFIDs to find the error and it increases the security as only benefit that user can access the IoT. The first Internet appliance was Coke Machine at Carnegie Mellon University. The machine automatically tells how many bottles are left and, moreover, how many are cold. The micro switches were attached which sense the coolness of bottle. The server was there to tell how many bottles are in which section and many other things about it.

3.3 Importance of IoT

Internet of Things is quite important in one's life as it helps us to live life smartly as well as it gives control to our lives. IoT provides business as many companies are working as 'IoT Service Providers'. IoT enables company to automate the working and reduces manpower which causes unemployment. Machines are far accurate than humans, so the output of the task is better than humans. The cost of the product decreases as machines do not require salary so the overall cost of manufacturing and delivering of product automatically decreases and provides transparency between the producer and consumer.

3.4 Future of IoT in India

Over the most recent two decades, India has developed as a centre point of data innovation. With regard to digitization or innovation progression, we have ascended to the top. With IoT quickly drawing closer, tech intellectuals anticipated that IoT market could be worth $1.7 trillion by 2020 [5], with more than 50 billion gadgets interfacing with the Internet of Things by that timeframe. Considerably Gartner anticipated that in 2017 there would be 8.4 billion associated gadgets to IoT around the world, 31% more from the earlier year. In any case, do you know from where will quite bits of that development come?

The United States like consistently stays at the cutting edge in innovative improvements; however, some tech aficionados are stating that India will be an excellent spot to search for IoT development. Some of them are saying that India will turn into the most significant shopper of IoT gadgets in the following 5 years. In any case, there are some who question the circumstances; however, one thing is basically evident that the eventual fate of IoT is brilliant in India, and on the off chance that you want to make a vocation in IoT, you can select yourself an IoT institute to help you in your mission of being an IoT master, a guaranteed IoT-proficient master or a full-stack IoT master.

3.5 Major Functioning Area of IoT

3.5.1 Weather Forecasting

There are various applications of IoT in various sectors like energy applications, healthcare applications, etc. IoT also has its huge application in weather forecasting.

No human on earth is aware of the power of nature and its uncertain changing behaviour. Humans have made many efforts to predict the nature of the earth and have switched from the traditional way of predicting weather to a more accurate and flexible way of prediction of weather with the help of IoT. Remote sensing technology is the method used for the real-time, accurate and more reliable analysis of weather [6].

3.5.2 Remote Sensing

Remote sensing is an activity used for obtaining information about an object using sensors and without having direct contact with the object itself. Remote sensing is classified into two types: active and passive remote sensing [7].

Remote sensors are divided into active and passive sensors. Sensors that have the ability to detect the energy when the natural energy is available are called passive sensors, whereas the sensors that have their own energy source are called active sensors. These sensors emit the radiation and direct it towards the target. The target also reflects back the radiation which is detected by these sensors.

- Example of passive remote sensor: Meteorological satellites
- Example of active remote sensor: Weather radars

3.5.3 *Meteorological Satellite*

Meteorological satellites are a good source of weather forecasting and warning services. They have a very huge coverage. Meteorological satellite images can predict the weather days before it comes to an area. They have a very basic application, i.e. identification of clouds, but apart from that they can also monitor cyclones, thunderstorms, floods, sandstorms, etc. They can also be a great source to observe volcanic eruptions [8]. In the meteorological satellite the low-cost sensors have been used for weather condition which may create disaster. These satellite images are observed (analysed) to identify the causes of uneven weather conditions and the future alert can be propagated.

3.5.4 *Radars*

Radar is a short form made for radio detection and ranging. It is now widely used in weather forecasting. It is active remote sensing equipment. Weather radars have application of measuring rain reflectivity. Radars can give finer details regarding weather than meteorological satellites. There are many more applications of radars in weather forecasting like monitoring cyclones and rainstorms. They are a great source of quantitative rainfall forecast.

3.5.5 *Communications*

Internet of Things is a network of interconnected devices over a network. Communication Technology

Many communication technologies are used such as Bluetooth, Wi-Fi, radio protocols, LTE-A, Zigbee, Mqtt, CoAP and RFID.

Communication models are as follows:

1. Request-response model
2. Subscriber-publisher model
3. Push-pull model
4. Exclusive pair model

IoT has been regarded as a third wave of communication after the Internet and mobile communication. Gateway like Zigbee plays an important role in communication as it makes integration sensor network with mobile communication.

Communication is the process to convey information from one place to another. Machines communicate with the help of some protocols. Some peripherals are UART (Universal Asynchronous Receiver/Transmitter).

In the recent years, there had been so much development in the field of IoT. In the end of this decade, nearly 20 billion devices will be connected through the IoT, and IoT market will grow up to 3 trillion dollars by 2026. But there are also some barriers in the development of IoT such as there are some security concerns as so many devices are connected.

3.5.6 Home Appliances

Nowadays systemic implementation of IoT is used for controlling the appliances of home and supervises the home globally; they can transmit with the devices through the Internet, with the help of Wi-Fi and Bluetooth [9, 26].

We can use Arduino, NodeMCU and Raspberry Pi as a server system by using mobile phones with the help of Wi-Fi. We can control the fan, washing machine, air conditioner (AC), refrigerator, lights, locks, water geyser and television. In automation, we can on or off anything without human involvement just like the above-given devices with the help of actuator. Today automation means a combination of the hardware and software system, and with the help of both things, the meaning of automation can be fulfilled.

Taking the aspects of the saving and conservation of energy, the technologies are nowadays of greater importance and increasing the notice for many researchers and countries.

Home can be automated to work faster; it can be made more efficient to save our time and energy. We can only use such devices or sensors or actuators through which we can make our house automated. Arduino can be used with some programming language like C, C++, etc. It depends on the user which programming language he/she can use. We can take an output by the machine according to user or yourself. For example, fixed an IR sensor to the door so that if anyone comes into the house, the door will open, the light will on, AC will on and windows and curtains will close. Another example related to automation of house that we can think of is that the door will be open only at 20° so only one person can allowed to enter at one time. With this kind of automation, information is sent to Arduino using a language and then Arduino gives the command to the actuators. We see that in this example the automated door can work; we also have done another thing in automated door, that is, we can fix a camera above the door so that if anyone is detected in front of the camera, the OTP is sent to the owner of the home and if the owner knows that person the owner sends the OTP to the Arduino and Arduino gives the confirmation to the actuator to open the door and the door will open. Hence automation can help the owner to provide the security from thefts and unknown person entry in the home. Another example we made is an automated dustbin. If we throw any waste material, the dustbin will be activated to collect the waste, and if it is full it can send a message to the sweeper to collect the waste and empty the dustbin.

We also make an automated AC. If the temperature is maintained, it can be off, and if we are going outside the home and we have no idea if the AC is turned off or not, we can switch off the main supply by the help of our mobile phone. And when water storage tank is not full, the pump will start filling up the water tank, and then when it is already full, the pump automatically is turned off. It is the best way to conserve light energy and also save water so we can also automate the main supply of AC or water pump.

Car garage can be automated. The shutter of the garage could be opened by the authorized users only, and to achieve this we can generate a pin for the user and then verify it using our hardware, and when the car is not in the garage, the shutter would automatically be closed, and this can be done by connecting the devices through the Wi-Fi or net connectivity. In October 2014 the idea of home automation came as a best thing as we cannot use new wiring to automate the home.

To make one's life easy and fast, we can use an IoT-based toaster so that the bread becomes ready as the user wakes up or after sometime.

The water heater should prepare warm water for bathing after the user has woken up, and it should be automatically switched on and be switched off when the certain temperature of the water is reached. It will also conserve energy.

The main advantage of this automation system over the manual work done by the human is quick response, accurate detection of emergency and accurate detection of unusual behaviour of the unauthorized user, leading to faster correction or diffusion in the critical situation.

3.5.7 Agriculture

Agriculture is the largest industry in the world, and when it comes to innovation here, there is lot more to explore. Indian agriculture can benefit a lot from an IoT-like solution which can help farmers and landowners to monitor the health of crop. Crops are sensitive to temperature, humidity, level of water, soil content and soil texture and most importantly the timing and quantity of pesticides.

The best use of IoT in Indian agriculture is for irrigation purpose. The quantity of water utilized for a particular field is measured so that available water can be used to its fullest. The start and stop of the motor can also be set, and it can be fully automated in accordance with the water required for the crop. Other IoT-based machines are also available for maintenance of the temperature, moisture, etc., but they are not widely used.

The major idea of Internet of Things is to make everything in this world work like a computer by connecting it with the Internet. When these things work like computers, then we named them as smart things. Nowadays IoT occupies a large sector of industries; almost in each and every device we can find the application of IoT. In hospitals, homes, even in agricultural field everywhere we are using IoT [10].

As population increases continuously, it is expected that global population will touch 9.6 billion by 2050 and it will be very difficult to feed this much population. Hence to tackle with this situation, many agriculture industries work on technology like IoT. With the help of IoT technologies, we can make smart devices like agricultural drones and process of using these devices named as smart farming.

Smart farming will enable farmers to reduce waste and help them to increase productivity range and quality. With the help of it, we will get a large amount of food and good quality of food as compared to ordinary farming. The following are ways to encourage smart farming:

3.5.8 Agricultural Drone

Technology has change time to time and agriculture drone is one of the best examples of this. Drones are used in agriculture to reduce labour work, to increase efficiency and to save time. Drones are used for spraying, crop monitoring, planting and field analysis.

3.5.9 Livestock Monitoring

Many farmers can use IoT to applications to collect data of their field, crop and health of their cattle. With this application these farmers are capable of identifying the health of their cattle if they are sick and then these cattle can be separated from the group of cattle to prevent others from spreading diseases. These applications help to identify disease in the cattle and inform farmers to cure it on a timely basis.

Benefits of Smart Farming

Various sensors lead to give proper data which on processing gives out proper input for crops.

1. Cost management and waste reduction.
2. Increase farming efficiency through process automation.
3. It increases product quality and volumes.

Application of IoT in Agriculture:

1. Precision farming
2. Agricultural drones
3. Livestock monitoring
4. Smart greenhouse

3.5.10 Smart Cities

IoT helps us in every field like in homes, cites, weather forecast, health, security, etc., and helps in making the things smart through automation. This is one of the major functions of IoT which helps in improving the infrastructure and environment, enhancing transportation system and optimizing the cost of public assets. As the world's population is increasing day by day, the urban areas are going through a lot of transformations like smart roads, traffic management, smart parking, smart waste management, etc. [11]. Here are some components of smart cities and their impacts on IoT are as follows:

3.5.11 Smart Parking

Smart parking involves the parking technique which does not require complicated infrastructure and is automated via IoT. In this type of parking system, sensors are present in the ground, and the driver receives signal via smartphone which allow them to park their vehicle in free space. This type of automation does not require high investment or high maintenance which ultimately helps in saving the time and improves the parking facility.

3.5.12 Smart Waste Management

Smart waste management system is very effective in managing waste as it helps in keeping the environment clean. Various types of sensors are used to monitor the dumpsters like ultrasonic sensors are used for measuring the level of dumpsters. Infrared (IR) and moisture sensors are used for separating dry and wet waste. If the container gets full, then the alert message is sent from the dumpster which prevents the overflow.

3.5.13 Smart Roads

Road can be termed as smart road if it is automated via smart equipment or simply via IoT technology whose ultimate goal is to make driving safer and greener. There are sensors which provide information to the driver regarding weather, traffic, accepted speed, etc. In case of some mishappening, the sensors sense the road environment and alarm the nearby police stations and hospitals accordingly. This type of road also generates energy to use street lights and charge electric vehicles.

3.5.14 Smart Railway Gate System

This system works with two infrared (IR) sensors for sensing the arrival and departure of railway engine whose ultimate goal is to make the railway gate system more reliable and automatic without human involvement. Humans can make mistakes, but technology cannot do any small mistake like humans. Human management of railway gate can lead to disastrous accident, so in order to make railway gate system more efficient, we can automate it via IoT. The work which the in-charge person does now can be done via smart equipment where sensors sense the train advent and actuators/motors associated with sensors open and close the gateways.

3.5.15 Smart Water Management

Management of water can be done via IoT technology, for example: controlling the occurrence of leakages and wastes via capacitive sensors, monitoring of the quality of tap water in the cities, controlling the swimming pool conditions remotely, detecting the liquid presence outside tanks and pressure variations along pipes, etc., are some basic applications which can be automated via IoT and can make the water management smart.

3.6 Toolkits to Implement IoT

3.6.1 TinyOS

TinyOS is an open-source operating system for wireless sensors. It uses nested C programming language. It is a useful OS which constitutes the sensors which require low power. Need of TinyOS comes into picture due to the problems associated with traditional OS. There are various TinyOS models like component-based model, event-based model, data model, thread model, programming model and network model. The features of TinyOS are that the processes are executed in a FIFO format meaning first in first out and process executes in a non-preemptive manner; it is completely non-blocking, and it is statically linked. There are vast applications of TinyOS in various fields of IoT and linked with all kinds of devices which depend on wireless sensor network. It is used to control environmental factors by measuring air pollution, forest fires by embedding a TinyOS in a small sensor which measures all pollution levels. Others examples of similar operating systems include Contiki, OpenWSN, FreeRTOS and RIOT [12].

Events-based design is also used by TinyOS which puts the CPU on rest when it is vacant or is not performing any task. An event can be anything like triggering of an alert on rising or falling of temperature over a certain value. The sensor motes will automatically go to sleep as soon as the event is over.

An application like smart factories and smart transits mostly uses a design like TinyOS, because to reduce energy requirements it is necessary to have very small memory footprint. TinyOS has been downloaded way over 35,000 by this time. Devices which use wireless sensor network lie in the application of TinyOS.

It is used in various fields such as the following:

1. *Machine Condition Monitoring*: There are many machines to machine applications which contain a lot of sensor interfaces. We cannot assign a complete computing environment according to each sensor that's why TinyOS is required by which we can perform energy management, security and sensor debugging.
2. *Smart Cities*: TinyOS is a suitable tool for the sensors of smart cities which uses low power, Internet facility, power grids and various other applications.
3. *Smart Vehicles*: Smart vehicles are independent and can be taken as a network of sensors. The sensors can communicate through wireless area network which requires low or less power; this makes TinyOS a perfect solution.
4. *Environmental Monitoring*: As we know that any TinyOS system can be easily placed in a small sensor that is why it is useful in air pollution monitoring, sensing of forest fires and also in preventing several hazardous natural disasters.

3.6.2 Mote Runner

A sensor network consists of a potentially large number of contiguously or geographically distributed autonomous devices, which can be called motes [13]. Motes feature many sensors and actuators which together work for monitoring and reacting to some physical or environmental conditions. The environmental conditions are temperature, pressure, sound or other like motion, etc. Other than sensors and actuators, after that each mote is generally equipped with a microcontroller, some transient and some persistent memory. It is also equipped with a wired or wireless communication device and a power supply, or we can say energy sources. The visualized size of a single mote hereby can vary; it can have the size as of a small coin or can have a size that of a shoebox. Its cost may also vary from a few dollars to a couple of hundreds of dollars. If the motes are mainly operated wirelessly, such network is called as a wireless sensor network (WSN).

Mote Runner which was created by IBM was designed to run on very small, standard and embedded controllers including low-power 8-bit processors, and because of which it reduces both investment cost at the initial stage and post-deployment and maintenance costs. It is primarily used as WSN that aims at tackling some of the problems related to good performance and cost management. We can do it in two ways; one is, it can happen directly, by providing a unified and virtualized execution environment so that we can get flexible configurations. And secondly it can also be done indirectly, by providing it an infrastructure and mechanisms so that we can implement higher-level and application domain-specified solutions.

From an engineering view, the Mote Runner's specifications are mainly designed so that it can be implemented portably and in a scalable and efficient ways building upon and carrying forward our experience with JCOP, the IBM JavaCard implementation. The Mote Runner implementation builds upon off-the-rack embedded (mote) hardware with a thin hardware abstraction layer which is written in 'C' and then an assembler that encapsulates any hardware-specific functionality. The next layer is also written in 'C' and it is a virtual machine; the run-time library is written in two languages which is 'C' and 'C#' and the 802.15.4 MAC layer is written in 'C'. Both the run-time library and the MAC layer open the APIs, so that application development can be done in higher-level languages like 'C#' and 'Java'. On top of the run-time library and the MAC layer, higher-level network communication protocol stacks are implemented or used, so that it can expose further APIs to the application programmers.

3.6.3 Contiki

Contiki is an open-source operating system which was developed for the IoT devices which have limited memory as well as limited processing power. Contiki was created in 2002 by Adam Dunkels, and now it has been developed worldwide by companies like Atmel, Cisco, Redwire, etc. [14]. It fits its entire operating system fewer than 30 KB space only.

Contiki provides low-power operations in system in which it works. This functionality of Contiki is called Contiki MAC. In other way we can say that it takes very less power to perform any operation in a system.

Contiki also provides a network simulator named Cooja which is a cross layer wireless sensor network simulator based on Java. It allows the simulation of network layers as well as emulation of hardware containing a set of sensor nodes.

Contiki programming model is based on protothreads, so that it can run efficiently in systems which has small memory. Protothread is mechanism of concurrent programming. It contains lightweight treads. It also provides multithreading.

Features of Contiki

- *Multitasking kernel* – Contiki provides the multitasking kernel, so that operating system can perform multiple tasks at a time.
- *Windowing system and GUI* – Contiki provides the windowing system. This software manages different parts of the screen separately.
- *Networked remote display* – Contiki provides the functionality to remotely control your system by using VNC.
- *Web browser* – Contiki provides a web browser that helps us access any information from the worldwide web.
- *Personal web server* – Contiki has its personal web server.
- *Screensaver* – Contiki has also the functionality of screensaver.

3.6.4 DeviceHive

It is an instrument/tool for IoT devices for communication and management purpose. It works in between the network of connected device and their server. It is an open-source technology and brings huge impact in M2M and IoT. DeviceHive is easily accessible. It also reduces the requirement to develop the messaging protocols and communication libraries bringing more focus on developing overall functionality for the project [15]. It includes a communication layer, control software and multi-platform libraries.

This platform covers the whole starting from data transition, validation and collection up to machine learning jobs and artificial intelligence. This basically offers monitoring tools, so it becomes easy to start discovery without being physically connected to the real hardware.

It provides its own GUI called as admin console. This will allow you build your own device, connect them to network and allow you to manage it. You can notify by recent updates and send instructions to connected device.

Protocol Available in DeviceHive

Generally, by default it provides REST or Websocket API and in addition MQTT protocol is available in DeviceHive. For using the complete service of REST, Swagger is provided by DeviceHive. Any device that supports REST API, Websocket or MQTT protocol can be connected to DeviceHive. The programming language used in its libraries are Java, Node.js, Python, etc.

Tools Integrated with DeviceHive

DeviceHive does the work in a different way. It allows you to develop your own devices. Because DeviceHive has its own GUI admin console, it also allows observing the recently uploaded messages/notification.

Customize the DeviceHive

As we know that DeviceHive is a open source-based system, if we want any changes, it is not easy. That's why we designed DeviceHive plug-in that subscribes to DeviceHive messages and implements required business.

3.6.5 Android as an IoT Tool

The Android Things is an extension of Android platform for the IoT devices. Android is designed for many things such as phones, tablets, watches and also cars. In Android Things we get multimedia connectivity and Wi-Fi and many things. It is ideal for powerful and intelligent device which provides securities [16].

Android Things is a new version of Android operating system that runs on our smartphones and smart devices. It is developed by Google, for the IoT development. Google released Android Things version 1.0 for the developers. It is the same like that which we are using in our smartphones, smart TV, smart watches, etc.

With the help of Android Things, we developed new commercial IoT devices which convert the way of living. It also helps us in spreading our business. It makes our life easy in a better way. Android Things provides us security which is more important nowadays, to secure our home, projects and many other things from the hackers.

Android Tools Used in IoT:

- Node-RED
- OpenHub
- RIOT
- Kinoma Create
- Eclipse IoT

3.7 Network Challenges in IoT

The IoT can be defined as the process or the management of the system in which there occurs data transfer or connectivity between the devices, objects and machines without having any interaction with humans, i.e just using the Internet for transferring data among them [17].

There is quite a great possibility that when this data transfer occurs among the devices, there may arise a problem about the connection between them in terms of network connectivity.

There are a number of challenges which can be seen in IoT related to network problem or connecting issues or problems in terms of transferring data between the devices.

Nowadays IoT devices are used everywhere for updating the technologies and to make the world smart. There are many applications of IoT devices in the real world, where we can use to make the task easy and smart such as smart city, smart home, smart health system, etc. There are many benefits of using IoT devices, but there are also some network challenges associated with it.

There are various challenges faced by developers while building IoT projects, but the main challenge is to send signal to other devices or servers. Not only sending correct data, but also the data should be in proper order, server or devices should have proper security and they should be properly managed and computed.

Some of the problems which are seen in terms of network challenges when developing an IoT application are listed below:

3.7.1 No Loss of Data

Suppose you are sending data from one device to another or to the server. You do not want your data to be lost and data should be in proper order. Nowadays in IoT projects, communication between devices or between device and a server is essential. It should be like real-time messaging.

Suppose you have two devices X and Y and a server. You want to send data from X to Y and then from Y to server. You need to be 100% sure that there should be no loss of data during communication. And you also need to pay attention that data is streaming to its correct destination; otherwise this will result to an error. For sending correct data to correct destination, we can use signalling. It helps in communication with the help of signals.

3.7.2 Security

Security is a huge challenge in IoT. What good will your device or server be if it gets hacked? As you all know if you are connected to the Internet, then you are vulnerable. You can be easily hacked. I mean what is the use of fingerprint or face recognition software on your house door if it can be unlocked by anyone. If we are sending data to device or to server, then it should be encrypted, which will make it harder to hack [18, 27]. We should provide proper authorization.

Connectivity of the devices is a very big issue; we have to make sure that there does not occur any loss of data during transferring among the devices. There are three aspects on which we have to look over it.

- *Authorization*: It is important for both server and user that they have proper authorization when sending or receiving the data streams during the transferring of data.
- *Open ports*: It is very vulnerable for an IoT device when it is communicating outside the given Internet protocol. We need a bidirectional communication for transferring the data over the server, but at any cost, we do not want any port to the Internet.
- *Encryption*: There should be an end-to-end encryption both at the server and user side so that the data may be protected from outside threats.

There must be such a system which identifies and authenticates devices, and then access control assigns it to their respective network segment automatically. Generally, the network from which IoT devices connect remains isolated from external activities.

Isolation helps prevent network from threats and attack. Similarly firewalls block compromised devices from connecting to the network with external command and control server.

3.7.3 Detect IoT Devices

Suppose your project is working fine and suddenly one of your devices gets crashed, then you should be able to identify which device is that. Or if your device goes offline or comes online, you need to know as well. You can achieve this by monitoring your devices [19]. For monitoring, you need real-time monitoring system and analytical reports. Suppose one of your device gets offline then your monitoring system will alert you.

Take an example if your company provides home security and suddenly motion sensors stop working, then you should be able identify the problem by monitoring them.

3.7.4 Power

Devices send a lot of data to other devices and to server, and this takes a toll on CPU. With this comes power drainage. With all this sending and receiving data, batteries drain and so you need low-power consumption. You cannot use all your power in carrying out a single task for a single device [20].

When thousands of devices are presently doing data transfer from server and vice versa, there may be a great possibility of battery drain and high-power consumption. At the same time, we cannot make our devices small and buy expensive CPU-embedded power system.

3.7.5 Bandwidth

Bandwidth is the maximum rate of data transfer. Not only power consumption is an issue but bandwidth consumption is also a major issue. We need a lightweight network that can send data to other devices or to server. And bandwidth of cellular network is expensive. Not everyone can afford this with many devices beacause the bandwidth consumption may be high. So bandwidth mau pay vital role in the communication of IoT devices,sensors and servers in communication [21].

We are living in a technical era where everything is going to be automated. From our house to offices everything is controlled by machines automatically. With this automation there are some issues rising like management, security and many more. Mostly all organizations are now adopting IoT for better functioning. They will need too many devices to do this (automation of organization); hence there is a possibility that most of those devices do not have such adequate security, which means they can be easily compromised to serve as slaves in IoT-based bonnets, become a medium for spread of malware or become a source to infiltrate networks.

Some attacks are Stuxnet, Mirai and Brickerbot. All had IoT component and caused widespread harm. These are just some of the popular attacks in IoT which we know of, and there are thousands of other attacks which are still going on. Most organizations have no efficient method to identify or trace these devices. The very first rule of security states that 'You can't protect or protect yourself against what you can't see'. The number of IoT devices and its applications and how much they connect to the network and the speed at which they are being implemented and which end user can connect is unprecedented. Most IT firms cannot tell you how many IoT devices are currently connected to their network. Some issues like where they are currently located and from which network they are connected and which resource they can access are difficult to find. Organizations having very strict policies have found that explosion of IoT in their networks overwhelmed their potential to trace these devices.

We know that while making an IoT-based model or system or project, there is not only one device which is connected over it. In some cases, the need for the devices may reach up to 100 per system. In such a scenario we have to take care of the bandwidth connection during the connectivity of the system. Bandwidth over a network comprising of cellular data is expensive and especially when there are 100 devices which are taking response or sending a response to the main server. For this, we need a lightweight network that can seamlessly transfer data from one device from others.

3.7.6 Signalling

When we connect any devices among each other, the flow of data takes place between them. It may happen that the connection between them is not reliable. In such cases, the use of data streams comes into play. It may happen that the devices are talking to the server or the devices are trying to establish connection between them only [22]. No matter what is the case, the transfer of data from point A to point B should reach quickly and be 100% reliable. While making any IoT system, we have to make sure of it.

3.7.7 Presence Detection

It is very important to know the status of all the devices that have been connected in an IoT system. The system administrator should know that when a device is taken off from the system or when new devices are being added to the system during the data transfer. It also gives relaxation to the administrator to monitor all the devices and rectify any error if it is caused in between them.

3.7.8 Role of Access Control

The effective strategy for IoT security is to ensure that you can see or track every device on network. Access control technologies require automatically recognizing such IoT devices, checking whether they are compromised and then providing controlled access to the secure region or end user. This can be done at digital speeds.

Another factor to be considered is location of such IoT devices [23]. Access control devices must able to find from where an IoT device is connecting or logging in the network.

Location-based policy is beneficial for such organization having branch offices. Access control technology should be able to synchronize with network and security controls to confirm the policy enforcement across network.

3.7.9 Access Control and Network Segmentation

There must be such a system which identifies and authenticates devices, and then access control assigns it to their respective network segment automatically. Generally network from which IoT devices connect remains isolated from external activities. Isolation helps prevent network from threats and attack. Similarly firewalls block compromised devices from connecting to the network with external command and control server. Access control system must be able to share device information to other security and management devices' information like they are collecting, provided they must be collected and correlated with other network resources. Such tools must do certain tasks like connect and surveillance IoT traffic so compromised devices can be identified and monitored by techniques like behavioural analysis and many more.

3.7.10 Access Control and Quarantining

Detection of malicious or unusual traffic from any IoT device automatically triggers an active response like shutting down that path of communication and by access control isolating it. One method is to divide the procedure in stages like evaluation, remediation and flush it [24].

Key Procedures

IoT devices are now very important especially in changing method of business and society. They are not only important for digital market but also for cybercrime activities. One of the major tasks is to protect organizations from such compromised IoT devices without affecting business agenda which is challenging. As the volume of such devices is increasing rapidly and traffic to network and their security, professionals also find it difficult to find 'such threat'. A small piece of such threat can have a major financial and reputational effect. An organization needs only an atmosphere where they can work securely. Access control must be able to minimize the impact on work by such threat or should we say negligence. Security ensures that visibility is established, access controls are universally applied, device intelligence is shared and rogue devices can be quickly removed with minimal impact to critical business transactions and workflows.

3.8 Threats and Security in IoT

Internet of Things is defined as the connection of ordinary devices to the Internet, such as light, fan, doors, etc. IoT is a branch of computer science. IoT improves the life of human beings and makes lifestyle more and more advanced.

Example of Internet of Things:

1. Many postal companies use tracking ID so that user can track their packages.
2. Nowadays printer gives option to buy new cartridges if ink is empty and also detects the amount of ink left in printer.
3. Smart home.
4. Smart medical facilities.
5. Family protection band.

Now the most important point in Internet of Things is security as latest devices are coming continually to this world. Internet of Things security focuses on protecting your Internet-enabled devices [25]. IoT security is the biggest concern of today and tomorrow.

Your connected devices or your smart devices require data, such as your full name, age gender, phone number, email ID, location, etc., which are all personal information that can help criminals in stealing your identity.

Security Measures to Secure Your Devices

- Always read the privacy policy of apps; beware how they are using your information.
- Do research when you buy an app. You should know what type of data to collect. Collecting data is not a bad thing, but it should be secure and not to be shared with third party.
- Use a VPN which helps to secure the data transmitted on your home or public Wi-Fi.

So a little caution can make your life better which is necessary for every network-based device network. Malware is one which needs to be focused while using the IoT networks. The details are as follows:

IoT Malware

As the number of IoT connected devices rises up, so the number of malware used to harm them. While traditional malware are encrypted to fully lock out the users from different devices and platforms, the malware attack could focus on limiting and disabling all the IoT devices functionally and stealing user's data at the same time.

For example, a simple IP camera is used to gather or capture the sensitive information using a vast range of locations, but the webcam can be locked and footage can be diverted from its path and easily leakage of information is possible.

3.9 Conclusion

IoT is a technology which has a great reach to the common people due to its use in day-to-day task. So this chapter focuses on the basic technology which controls and enables the IoT devices. The toolkits are used to develop IoT application, their connectivity and controlling the network of IoT devices. The future research direction can be done in the area of controlling IoT devices through mobile applications. The easier interface and tools can be developed which will lead to more innovation in this area. The security in the communication network and devices can be optimized as a future research direction. The sensor technology is a backbone of IoT so sensors are considered more vulnerable. So more security must be needed to make them trustworthy for users.

References

1. Boeckl, K., Boeckl, K., Fagan, M., Fisher, W., Lefkovitz, N., Megas, K. N., & Scarfone, K. (2019). *Considerations for managing Internet of Things (IoT) cybersecurity and privacy risks*. Gaithersburg: US Department of Commerce, National Institute of Standards and Technology.
2. Zhang, Y., Peng, L., Sun, Y., & Lu, H. (2018). Intelligent industrial IoT integration with cognitive computing. *Mobile Networks and Applications, 23*(2), 185–187.
3. Pandey, N. K., Chaudhary, S., & Joshi, N. K. (2017). Extended multi queue job scheduling in cloud. *International Journal of Computer Science and Information Security (IJCSIS), 15*(11), 1–8.
4. Gershenfeld, N. A., & Gershenfeld, N. (2000). *When things start to think*. New York: Macmillan Publishers.
5. Gubbi, J., Buyya, R., Marusic, S., & Palaniswami, M. (2013). Internet of things (IoT): A vision, architectural elements, and future directions. *Future Generation Computer Systems, 29*(7), 1645–1660.
6. Wang, B., Kong, W., Guan, H., & Xiong, N. N. (2019). *Air quality forcasting based on gated recurrent long short term memory model in internet of things*. IEEE Access.
7. Haider, I., Höberl, M., & Rinner, B. (2016, May). Trusted sensors for participatory sensing and iot applications based on physically unclonable functions. In *Proceedings of the 2nd ACM international workshop on IoT privacy, trust, and security* (pp. 14–21). Xi'an, China: ACM.
8. Chapman, L. (2015, July). Urban meteorological networks: an urban climatologists panacea. In *9th International conference on urban climate (IAUC & AMS), open plenary speech* , Météo France (pp. 20–24). hal
9. Hasan, M., Haque, M., & Rabbi, M. (2019). Design and implementation of voice controlled home automation system. Daffodil International University Dhaka, Bangladesh. https://hdl.handle.net/123456789/2402
10. Gondchawar, N., & Kawitkar, R. S. (2016). IoT based smart agriculture. *International Journal of advanced research in Computer and Communication Engineering, 5*(6), 838–842.
11. Liu, Y., Yang, C., Jiang, L., Xie, S., & Zhang, Y. (2019). Intelligent edge computing for IoT-based energy management in smart cities. *IEEE Network, 33*(2), 111–117.
12. Berger, A., Ambe, A. H., Soro, A., De Roeck, D., & Brereton, M. (2019, June). The stories people tell about the home through IoT toolkits. In *Proceedings of the 2019 on designing interactive systems conference* (pp. 7–19). San Diego, CA, USA: ACM.

13. Balasubramaniyan, C., & Manivannan, D. (2016). IoT enabled air quality monitoring system (AQMS) using raspberry Pi. *Indian Journal of Science and Technology, 9*(39), 1–6.
14. Durmaz, C., Challenger, M., Dagdeviren, O., & Kardas, G. (2017). Modelling contiki-based IoT systems. In *6th Symposium on languages, applications and technologies* (*SLATE 2017*). Schloss Dagstuhl-Leibniz-Zentrum fuer Informatik.
15. da Cruz, M. A., Rodrigues, J. J., Sangaiah, A. K., Al-Muhtadi, J., & Korotaev, V. (2018). Performance evaluation of IoT middleware. *Journal of Network and Computer Applications, 109*, 53–65.
16. Alghamdi, K., Alqazzaz, A., Liu, A., & Ming, H. (2018, March). Iotverif: An automated tool to verify ssl/tls certificate validation in android mqtt client applications. In *Proceedings of the Eighth ACM Conference on data and application security and privacy* (pp. 95–102). Tempe, AZ, USA: ACM.
17. Kumar, N. M., & Mallick, P. K. (2018). Blockchain technology for security issues and challenges in IoT. *Procedia Computer Science, 132*, 1815–1823.
18. Ammar, M., Russello, G., & Crispo, B. (2018). Internet of things: A survey on the security of IoT frameworks. *Journal of Information Security and Applications, 38*, 8–27.
19. Guo, H., & Heidemann, J. (2018, August). Ip-based IoT device detection. In *Proceedings of the 2018 workshop on IoT security and privacy* (pp. 36–42). Budapest, Hungary: ACM.
20. Martinez, B., Monton, M., Vilajosana, I., & Prades, J. D. (2015). The power of models: Modeling power consumption for IoT devices. *IEEE Sensors Journal, 15*(10), 5777–5789.
21. Pustišek, M., Dolenc, D., & Kos, A. (2019). LDAF: Low-bandwidth distributed applications framework in a use case of blockchain-enabled IoT devices. *Sensors, 19*(10), 2337.
22. Xu, T., Liu, F., Li, A., Masouros, C., & Darwazeh, I. (2019, April). Constructive interference precoding for reliable non-orthogonal IoT signaling. In *IEEE INFOCOM 2019-IEEE Conference on computer communications workshops* (*INFOCOM WKSHPS*) (pp. 590–595). Paris, France: IEEE.
23. Qiu, J., Tian, Z., Du, C., Zuo, Q., Su, S., & Fang, B. (2020). A survey on access control in the age of internet of things. *IEEE Internet of Things Journal, 7*(6), 4682–4696.
24. Candal-Ventureira, D., Fondo-Ferreiro, P., Gil-Castiñeira, F., & González-Castaño, F. J. (2020). Quarantining malicious IoT devices in intelligent sliced mobile networks. *Sensors, 20*(18), 5054.
25. Siddiqui, S. T., Alam, S., Ahmad, R., & Shuaib, M. (2020). Security threats, attacks, and possible countermeasures in internet of things. In *Advances in data and information sciences* (pp. 35–46). Singapore: Springer.
26. Kumar, A., Srinivas Kumar, P., & Agarwal R. (2019). A face recognition method in the IoT for security appliances in smart homes, offices and cities. In *2019 3rd International Conference on Computing Methodologies and Communication* (*ICCMC*). (pp. 964–968). Tamilnadu, India: IEEE.
27. Bhushan, D., & Agrawal, R. (2020). Security challenges for designing wearable and IoT solutions. In *A handbook of internet of things in biomedical and cyber physical system* (pp. 109–138). Cham: Springer.

Chapter 4
Smart Self-Immolation Prediction Techniques: An Analytical Study for Predicting Suicidal Tendencies Using Machine Learning Algorithms

Kaushik Chanda, Ahona Ghosh, Sharmistha Dey, Rajesh Bose, and Sandip Roy

4.1 Introduction

In the present situation, depression and suicidal attempts have been increasing in a surprisingly high rate. According to a report published by WHO, every year 264 million of people face the problems of depression and it is increasing day by day.

During this COVID 19 situation, the increasing rate of depression is really a cause of worries. Not only the current situation, depression is an eye-raising situation since the last few years. According to a report published by WHO, since the last 45 years, the suicide rate increased by 45% around the world [1].

So, to support that different angle, this chapter has been written focusing on prediction of suicidal attempt through advanced technologies. In this smart city era, the new age technologies can also facilitate our lifestyle monitoring by several machine learning or data analytics algorithms used for prediction of an unfortunate occurrence of suicide.

It has been observed that among major causes of suicide, depression is one prime factor. In the next segment, the role of depression behind a suicidal attempt has been discussed.

4.1.1 Depression as a Major Cause of Suicidal Attempt

Depression is a common illness in today's world. Almost 264 million people worldwide are affected with some form of psychological disorder [2]. It should not be mixed up with usual mood fluctuations, empty feelings or short-lived emotional

K. Chanda (✉) · A. Ghosh · S. Dey · R. Bose · S. Roy
Department of Computational Science, Brainware University, Kolkata, India
e-mail: kc.cs@brainwareuniversity.ac.in

M. Moh et al. (eds.), *Smart IoT for Research and Industry*, EAI/Springer Innovations in Communication and Computing,
https://doi.org/10.1007/978-3-030-71485-7_4

outbursts that occur as a response to the challenges faced in our daily lives, but when these occurrences are durable and have simple intensity, it may become a serious health issue. This may result in significant mental and physical sufferings for the affected person and that may lead to function poor in professional as well as personal life. If depression is left untreated for long, then it often leads to suicidal thoughts or attempts. Almost 810,000 people expire every year for the reason of suicide which is caused by extreme depression. Suicidal death is one of the prime reasons of death in teenagers as well as adults below the age of 30 [2, 31].

Although depression has high severity, it is one of the most curable diseases. Effective treatments for depression do exist, but several barriers to effective mental healthcare make it quite inaccessible for almost 76–85% of lower-income and middle-income families from economically backward countries, even if such countries have smart cities and technologies [36]. Several existing barriers to effective mental healthcare include a lack of resources, lack of properly trained medical facility providers and the social disgrace related with mental disorders. Another problem for this mental health treatment is mobility problem [32, 38]. Internet-based assistance of primary physicians or doctors may be able to treat a larger number of patients, and the waiting queue for psychological treatment can be minimized in this way [35]. It has been found by a survey done in Australia that suicide frequencies are more in village areas than in towns, and fewer people in rural locations—mainly young men—search for help having psychological ailments [33, 34]. Another barrier to actual care is inaccurate medical assessment. For high-income and middle-income groups, it is lack of proper training of people and lack of proper diagnosis, which in turn generates false-positive cases in a large number.

A World Health Assembly resolution passed in May 2013 called for a comprehensive, coordinated response from all countries towards addressing and finding solutions to the problems caused due to mental health issues [2]. There are several categories and symptoms of depression in the following section to make a clear understanding about depression as a prime cause of suicides around the world in the last few years.

4.1.2 Categories and Indications of Depression

Psychological disorder can be divided into three different categories depending on the number and severity of indications. The categories are mild, moderate or severe. A person having mild depression may experience some struggle in continuing their work and maintaining their social life, whereas a person suffering from severe depression might cease to function completely and start abusing drugs or start to show suicidal tendencies. Based on their behaviour, depressive disorder may be divided into these following two categories [3, 4]:

- *Recurrent depressive disorder*: A person suffering from this type of depression experiences repeated depressive symptoms. The affected person shows a general loss of interest, lack of energy for several weeks, insomnia or abnormal sleep cycles, etc.

- *Bipolar affective disorder*: A person having bipolar affective depression experiences both depressive and maniac symptoms, quite fluctuated from normal behaviour. The kind of depressive disorder exhibited by the person includes hyperactivity, inflated self-esteem, sudden urges to eat a lot, difficulty in speech and decreased need for sleep.

4.1.3 Contributing Factors and Prevention of Depression

Depression is caused due to a complex combination of social, biological, socio-economical or psychological factors. People going through adverse life events (psychological or physical trauma, marital violence, a family member's death, unemployment, etc.) are more likely to develop depression. Depression can lead to anxiety which again increases the level of depression, just like a vicious cycle. Several other factors such as some social media platforms may act as a catalyst to depression.

It has also been observed that physical health and depression are interrelated. For example, any cardio problem can cause depression and the other way around.

It has been witnessed that programs related to optimistic activities very often reduce depression.

4.1.4 Need of Proper Analysis and Diagnosis for Depression to Prevent Suicide

Though alarming increase in mental health problems in recent times has become a matter of concern to all of us, treatment is neither very expensive nor impossible to do. The only concern is proper and timely diagnosis. Minor to moderate depression can be treated with psychological therapies, and some of the available psychological treatments are behavioural activation, interpersonal psychotherapy and cognitive behavioural therapy [40]. In cases of severe depression, both psychological and antidepressant-based treatment methods [41], like tricyclic antidepressants (TCAs) and serotonin reuptake inhibitors (SSRIs), are used. Healthcare providers should remind that prescribed antidepressants may have adverse effects and, in some cases, result in the affected person getting addicted to the antidepressant. Antidepressants must not be used for the treatment of depression in children and should not be considered as the primary type of treatment in adolescents, and even if they are to be used, it should be done with extra caution [39].

4.1.5 Impact of Depression

Depression and mental disorders associated with it can have a very negative effect on all aspects of an affected individual's life. It can dramatically reduce productivity and effectiveness, ruin relationships with family and friends and reduce the affected person's self-esteem [42] and confidence by such a degree that renders them unable to participate within the community. Research has shown that mental health is strongly related to the physical health of an individual and vice versa, so when a person is depressed for a significant amount of time it might cause cardiovascular diseases, constant headaches, loss of appetite, etc. Most suicide cases are caused due to depression. Depression affects all genders and every section of the society, young and old, rich and poor of every single country.

4.1.6 Signs of Depression

Depressed people typically have subsequent symptoms:

- Loss of enjoyment in most activities
- Constantly feeling tired or lethargic
- Hypersensitivity and emotional vulnerability
- Being easily distracted, inattentive and forgetful
- Inconsistent sleep patterns, insomnia
- Appetite and weight changes
- Very low sex drive
- Very low self-esteem, constant feelings of guilt
- Constant feelings of self-blame, feeling like an outcast, paranoia
- Being pessimistic about the future
- Persistent thoughts of suicide
- Alcohol or drug consumption
- Being angry and irritable
- Physical pain or health disorders (e.g. headache, digestive problems)

4.1.7 Graphical Representation of Worldwide Suicide Record (Age, Country and Gender-wise)

Worldwide suicide records from 1985 to 2015 from a public domain dataset [1] have been analysed and simulated using R-3.6.3. Depression is a prevalent mental disorder which often leads to suicidal ideation and attempt. The rate of worldwide suicide increase from 1985 to 2015 is alarming as shown in Fig. 4.1a, and 75+ aged people are prone to commit suicide than the young generation. Figure 4.1b shows

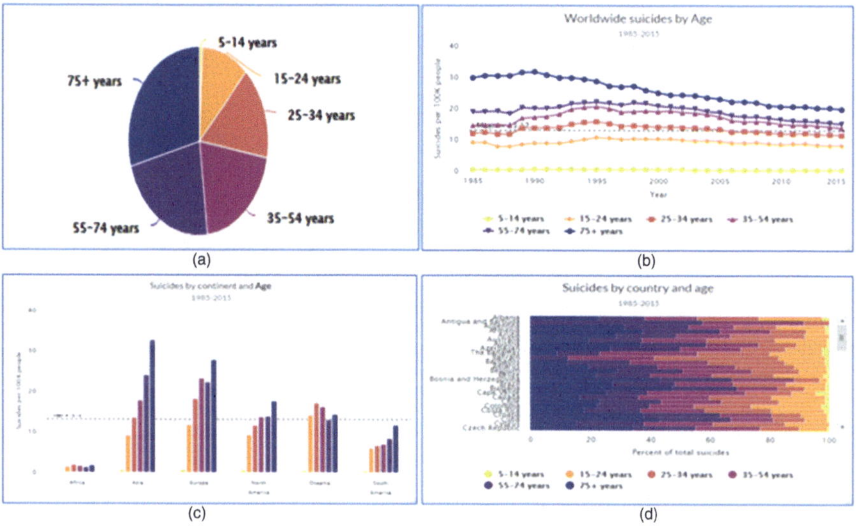

Fig. 4.1 Age-wise suicide analysis 1985–2015. (**a**) Worldwide suicide by age, (**b**) worldwide suicide by age from 1985 to 2015, (**c**) suicides by continent and age (1985–2015), (**d**) suicides by country and age (1985–2015)

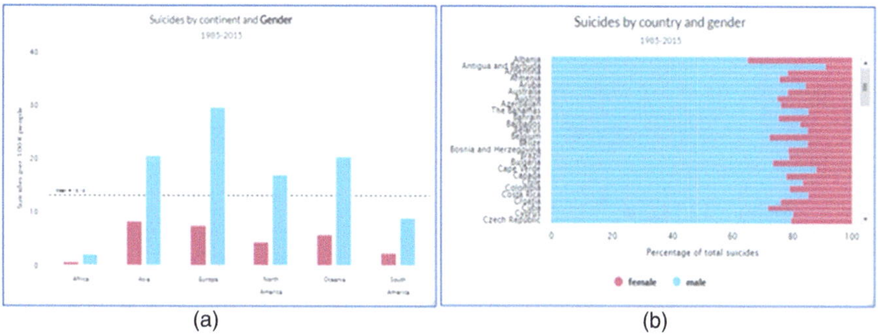

Fig. 4.2 Gender-wise suicide analysis 1985–2015. (**a**) Suicides by continent and gender (1985–2015), (**b**) suicides by country and gender (1985–2015)

that the suicide rate commitment has diminished in the people of age group 75+ years whereas it remained almost same in case of people having age group 15–24 and 25–34 years. According to Fig. 4.1c, the suicide rate is highest in Asia and lowest in Africa. Figure 4.1d shows country-wise suicide rate from 1985 to 2015.

The gender-wise suicide analysis depicted in Fig. 4.2 shows that rate of suicide of male is higher than that of female worldwide.

Suicide rate by worldwide countries is shown in Fig. 4.3, and suicide rate by years in Fig. 4.4 shows that the rate was highest in 1995, and the number of suicides has been a little bit controlled in the later phase.

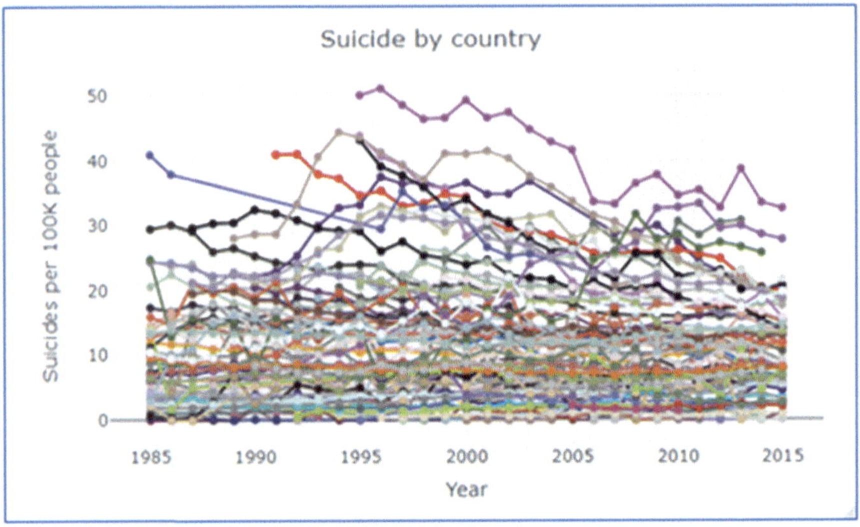

Fig. 4.3 Suicides by country

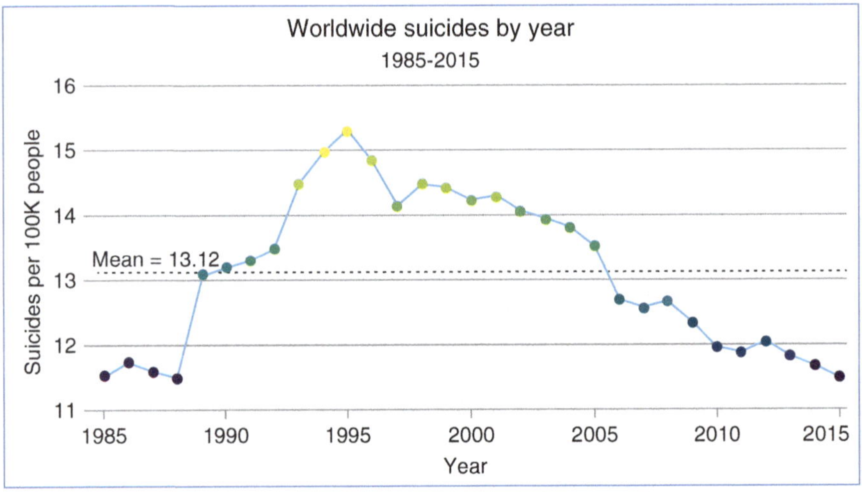

Fig. 4.4 Worldwide suicides by year (1985–2015)

4.2 Related Background Study

Suicide is a fatal issue causing the loss of many lives from every corner of the world, and according to World Health Organization, every year almost 8 million people lost their lives and the 70% of the death rates are from lower-income and middle-income countries [1]. By 2030, in every 20 seconds, one life may be destroyed forever [2]. To outline the recent trends in suicides and learning its risk factors in India, multiple correspondence analysis (MCA) and subset MCA [3] have been used which studies the association of attributes like gender, age group and profession. Several researchers have performed study on different supervised [8, 15–17, 27, 28] as well as non-supervised [25, 44–46, 52] ML algorithms for depression analysis. Some researchers have taken the help of data analytics techniques like principal component analysis or other predictive analytics. Such study has been performed on India also [52], and the study predicts the suicidal trends in India, but this study was not independent of age group. Among many causes, depression is one important reason of suicide. Prediction and prevention of suicide ideation in modern world is one of the biggest concerns because of its effect in untimely death of different age group people specially in young generation. Several research works have been undergone, and different systems have been implemented in this context [21, 51–54, 59, 60]. Some researchers have used supervised learning algorithms for predicting suicidal tendency, whereas some have applied unsupervised learning methods to predict and prevent suicide. Ji et al. have reviewed applications of different machine learning methods in suicidal idea detection and prevention [16] where data deficiency, annotation bias, lack of proper understanding the intention and data imbalance have been identified as the limitations of the existing approaches to suicide prediction. Depending upon the machine learning methods used to detect suicide, we have divided our analysis into following three categories, i.e. supervised learning, unsupervised learning and reinforcement learning as shown in Fig. 4.5. Based on the field of applications, our analysis has been divided into four categories, i.e. questionnaire analysis, clinical interview of the suicide ideates, analysis of social media post and suicide note text. In case of India, the predictive analytics for suicidal death has been performed by Amin and Syed [52]. In their study they have provided a clear picture of India's suicide scenario. They have adapted ML techniques like artificial neural network or support vector machine. Their study is good, but it was incapable of independently tracking suicidal attempt according to the age group. Scientist John Torous and his coresearchers have focused on an area where a study has been provided to understand suicidal tendencies using smartphones and improved sensors. But the main drawback of their study is the lack of appropriate data measurement techniques [58]. Table 4.1 discusses several related researches done in this area:

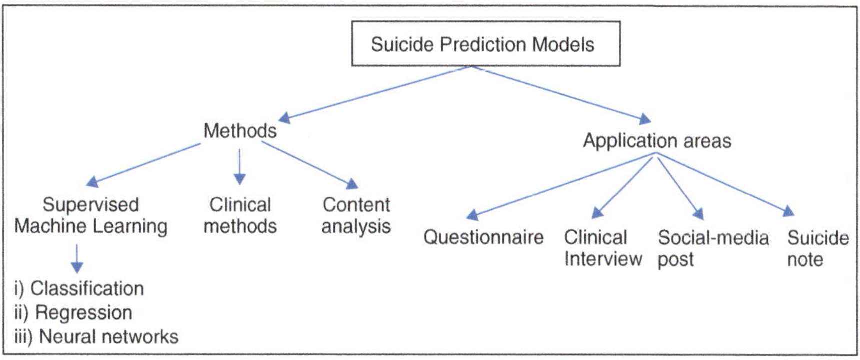

Fig. 4.5 The categorization of suicide prediction and prevention: methods and applications

Table 4.1 Comparative analysis of the existing methodologies for suicide detection and prevention using machine learning

Ref. no.	Objective	Methodology used	ML algorithm used	Performance	Drawback(s)
[4]	To study use of ML algorithms in depression analysis	Syntax analysis of Twitter and Weibo posts by different age groups to find certain keywords and their relevance Semantic analysis to find the general emotion of paragraph Classification of posts according to the symptoms of depression	SVM	6.09% of relevant Weibo posts reflected symbols of anxiety towards depression	Requirement of social media post data verification to achieve perfection in prediction
[5]	To identify suicide-prone individuals with specialized crisis management including direct message assessment and one-to-one counselling	Annotation of social media posts on suicide note for training dataset Automatic identification of posts involving suicidal thought and behaviour	SVM, DT, RF, LR with tenfold cross-validation	(i) SVM Precision 0.87 Accuracy 0.86 F-measure 0.84 Recall 0.81 (ii) DT Precision 0.79 Accuracy 0.78 F-measure 0.75 Recall 0.73 (iii) RF Precision 0.86 Accuracy 0.82 F-measure 0.79 Recall 0.74	Consideration of only a microblog group, no access to media platforms like school bulletin board, online suicide group, online self-help group No consideration of factors like posting time, frequency, etc. other than the text

(continued)

Table 4.1 (continued)

Ref. no.	Objective	Methodology used	ML algorithm used	Performance	Drawback(s)
[6]	To predict the suicide ideation and behaviour in some former and current Canadian armed force	Extraction of critical variables from Patient Health Questionnaire (PHQ) to diagnose the risk of suicide ideation Intervention for early prevention to improve health and well-being	Minimum redundancy maximum relevance (mRMR) feature selection, RF and fivefold cross-validation	AUC score 84.4% ± 4.4 with 25 variables and 81% ± 2.7 with only ten variables	Not mentioned
[7]	To predict the suicide ideation in a selected population	Random selection of individuals from non-suicide ideators to avoid class imbalance problem Classification of testing dataset to predict the suicide ideation	RF in training and elimination of recursive feature via tenfold cross-validation	Accuracy in test set 78.3%, Accuracy in total population 82%, Sensitivity 77%, Specificity 79.2%	Requirement of more study of public health and clinical data, biomarkers to predict more critical suicide risk like self-harm and suicide attempt
[9]	Detection of suicide attempters among suicide ideators	Resampling of suicide ideators by Synthetic Minority Oversampling Technique (SMOTE) to obtain data for person who attempted suicide and also non-attempted	RF in training and recursive feature elimination via tenfold cross-validation	Accuracy 88.9%	Dataset contains very simple scale and questions, no comparison with other algorithms. Use of only class balanced data, no actual data with bias class ratio
[10]	Design a pre-processing method to improve prognosis possibilities for suicide tendency prediction	Pre-processing Feature selection with PCA and evolutionary search	Principal component analysis and support vector m	Performance with MLP 92.18%, SVM 76.56%, HMM 76.56%, RBF classifier 86%, RBF network 84.3%, RF 85.93%	Experiment with limited number of classifiers, feature selection and pre-processing methods

(continued)

Table 4.1 (continued)

Ref. no.	Objective	Methodology used	ML algorithm used	Performance	Drawback(s)
[11]	To detect suicide risk and predict future suicide attempts	Classification by random forest for statistical modelling	Anatomical Therapeutic Chemical classification, Level V (ATC, Level 5)	Recall 95% Precision 77%	No integration of potential precipitating events like job loss Need experiment with larger dataset
[19]	Design of Twitter-based event detection for suicide	System developed based on PHP, MySQL and Twitter API	Text mining	Not mentioned	Lack of incorporation of public tweet
[25]	Suicide prevention by feature extraction	Medical data modelling Tested with more than 100 occurrences	Feature extraction with linear model K-nearest neighbour algorithm	High-risk performance is poor (34%) Moderate-risk performance is 45%	Earning trust from the medical practitioners is a challenge Probability of matching between manual strategy and ML strategy
[26]	Randomized machine learning technique	They test five machine learning techniques with three feature sets	Sparse logistics regression and decision tree	AUC 74%	Use of single retrospective cohort Use of randomized method for complex dataset
[27]	Mobile application for suicide prediction	e-Health application Aapproach for intelligent health (i-health) application for development Knowledge execution and application of data mining technique	Knowledge execution Unsupervised clustering	Not mentioned	Use of randomized data for complex dataset

(continued)

Table 4.1 (continued)

Ref. no.	Objective	Methodology used	ML algorithm used	Performance	Drawback(s)
[28]	To predict individual risk of suicide attempting	Dataset was from at least five hospitals. They have performed computation using python with tensor flow	Neural network model	Comparatively low accuracy rate	Data privacy issues
[29]	Suicide prediction through sentiment analysis	They used Weka data mining tool Information collected from Twitter Twitter4J API has been used for data collection	Support vector machines (SVM), naive Bayes algorithm and maximum entropy	Their result can predict suicidal ideation using Twitter data	No experiment with multilingual database
[30]	To study the impact of AI in predicting suicide outbreak and study the behaviour	Literature survey paper OVID Medline, EMBASE and PsycINFO, these three databases were used	Survey paper, no implementation	With ML achieving high accuracy is possible (85%–94%)	Should be used together with clinical expert report/data
[33]	To measure suicide rate as per the demographic Prevalence of depression	They have done survey and used data analytics	Machine learning techniques	In rural areas suicide rate are higher in men than in women In metropolitan area suicide rate is 31.8 per 100 000	The study is mainly based in Australia
[43]	Study the relationship between physical work and depression	Review work to study the relation of physical activities with depression	No specific algorithm	Established hypothetical relationship between five physical activities and depression	Only five types of physical activities have been discussed, more issues are needed to be discussed Frequency and volume of activity should be included

(continued)

Table 4.1 (continued)

Ref. no.	Objective	Methodology used	ML algorithm used	Performance	Drawback(s)
[44]	Twitter text analysis with the help of ML algorithm	Text mining and analyse with intelligent ML algorithms	Support vector machine and neural network	They have attained 95.2% accuracy using SVM and 97.6% using neural network	No focus on other platform except Twitter Only English has been considered Data is very limited as actual suicide note in social media is limited
[46]	To investigate clinical scale for identifying actual suicide attempt	Measuring psychiatric rating scale	Artificial neural network	The model shows overall 93.7% accuracy for 1 month, 90.8% for 1 year, and 87.4% accuracy in lifetime suicide attempt detection	Only five self-report scales have been used
[52]	To study prediction of suicide attempt in India using machine learning algorithms	Survey to predict suicidal attempt in India	Artificial neural network, support vector machine	77.5% estimation accuracy using ANN 81.5% accuracy using SVM	Independent tracking of suicidal attempt irrespective of any age group
[55]	To study a person according to age and strategy and find how to improve	Measuring suicidal tendencies of a person	Random forest classifier	83.7%	Subjective Data Insufficient data to cover original factors Overfitting problem has not been analysed
[58]	To understand innovative approaches for better understanding of suicide	To make a study among several ML techniques	Big data analytics and ML techniques	Survey paper	Lack of appropriate data collection mechanism

4.2.1 *Study of Several Machine Learning Models*

Machine Learning Methods Early detection of suicidal thoughts can be an effective way of suicide prevention. This section investigates and analyses different machine learning techniques applied to suicide prediction and prevention systems. In this section a comparative study among several machine learning algorithms has been provided. The below figure contains a hierarchical classification of machine learning algorithms.

4.2.1.1 Supervised Learning

Most of the researchers have implemented classification and regression-type supervised learning methods like support vector machine (SVM)[5, 10, 29], random forest (RF) [52], decision tree algorithm, logistic regression (LR), multilayer perceptron (MLP), radial basis function classifier and hidden Markov model (HMM) to predict suicide ideation and prevent the victim from such a fatal attempt. Ramalingam et al. predicted the study on depression analysis using SVM. They have done their work using emotion detection, which can predict depression level within a person [4]. People nowadays are addicted to social media. Suicidal tendency can be predicted from social media status also. Some researchers have presented their prediction based on social media data mining. Currently Facebook also has developed its own pattern recognition algorithm which monitors the posts by different types of users for probability of suicide and connects them with the concerned ones if applicable. Proactive suicide prevention method is introduced by Liu et al. Tadesse et al. [8] have worked on social media data. They have approached early detection of suicidal prediction using deep learning. After collecting several phrases from the social media, application of convolutional neural network and LSTM (long short-term memory) has achieved more than 90% accuracy. Prediction of reasons behind suicide in the future is the main aim of Amin et al. [15] where analsis of the pattern of the recorded suicide cases has been carried out at first, and then using SVM and neural network, the future trend has been predicted. SVM achieved 81.5% accuracy, whereas neural networks have given 77.5% accurate results in classification [17]. Binary classification methods have also been used for predicting suicidal tendency [24, 28]. Harish Bhat and Sidra Mellor worked in this area. They have observed five hospital data and they implied neural network model to analyse the data.

Some researchers have worked on unsupervised learning. S. Berrouiguet and his coresearchers have worked upon intelligent e-health application for suicide prediction, and they have used unsupervised clustering technique [27]. They have designed a mobile-based application and by merging electronic health record and electronic momentary assessment data for users, they tried to find knowledge extraction and to predict the analysis [61].

4.2.2 Clinical or Medical Methods

Consultation with psychologists is required whenever someone feels demotivated or depressed for gaining back the self-confidence, and by some therapy, he/she can get back to have a normal healthy life, and several clinical methods are there for suicidal risk assessment and prevention at an early stage [32, 33, 43].

4.2.3 Content Analysis

Different posts in social media reveal a user's choice of language, lifestyle and many more information about the user. By exploratory analysis of data, the linguistic clues of suicide ideators and attempters can be defined, and keywords or phrases, like 'lonely', 'kill', and 'cutting myself' which are related to suicide, can be identified. Strong negative feelings like hopelessness and anxiety can lead to suicidal thoughts. Suicide-related dictionary has been built in this context [4, 19, 37], and several researches are being carried out on analysis of social media content.

Researchers have also performed a study analysing the rate of suicide according to demographic data [33], the suicidal tendency according to their behaviour [32] or the study of relationship of physical activities with depression [43].

4.3 Popular Machine Learning Algorithms Used in Suicide Prediction and Detection

There are many popular ML algorithms used for prediction of suicidal tendency. Most of them are supervised algorithms [26, 27, 42], because labels are assigned to the training dataset, and following those class labels, the testing is performed.

4.3.1 Pre-processing Algorithms

Data pre-processing, the first step of creating a machine learning model, is the process of making the raw data suitable for its corresponding machine learning model according to the requirement. Most of the researchers have predicted and tried to prevent suicide by analysing depressed person's social media posts and different types of posts in social media.

4.3.2 Classification Algorithms Used for Suicidal Tendency Prediction

Classification is a supervised learning method where the main aim is to categorize a given dataset into classes, often called as label, target and category also. In our concerned domain, classification has been mostly used [12–14, 18, 20] to classify the suicide ideators and suicide attempters and sometimes suicide ideators and non-suicide ideators. Some popular algorithms applied to predict and prevent suicide in existing literature are discussed below.

4.3.2.1 Random Forest Algorithm

It is a supervised algorithm which is used for classification as well as regression. Working procedure of random forest algorithm (as shown in Fig. 4.6):

Step 1: Select random data sample from defined dataset.
Step 2: The algorithm will produce a decision tree for each sample in the dataset.
Step 3: For each decision tree, a prediction result will be generated.
Step 4: Voting of the target sample will be achieved for every prophesied result.
Step 5: Choose that value which has got a maximum vote as the predicted final result.

The diagram of random forest algorithm has been depicted in the figure below (Fig 4.6).

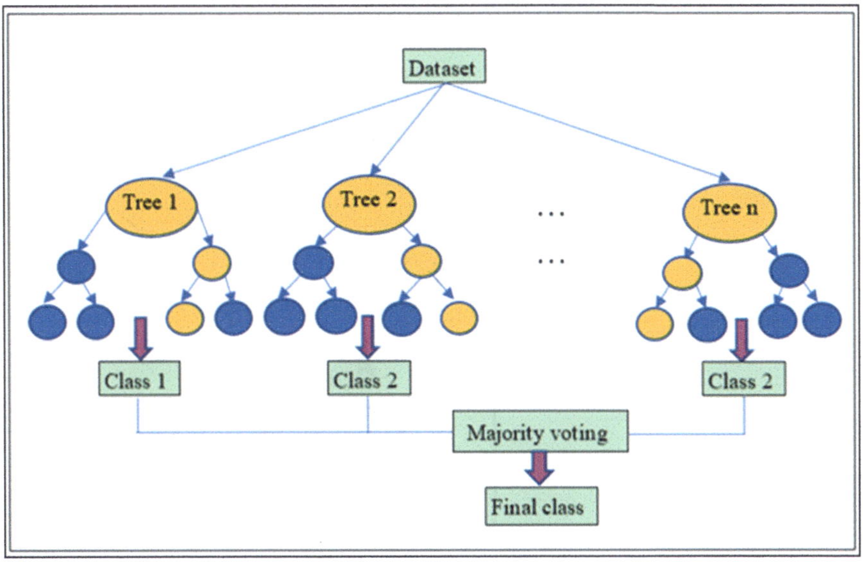

Fig. 4.6 Working mechanism of random forest

4.3.2.2 Support Vector Machine (SVM)

It is a supervised machine learning algorithm that can be implemented in classification as well as in regression, but it is most popular as a binary classifier in the existing works of suicide ideation detection and prevention, which classifies the new data points using a hyperplane separating the class 1 and class 2 data points. The working mechanism of SVM is shown in Fig. 4.7.

4.3.2.3 k-Nearest Neighbour (k-NN)

The k-NN algorithm in machine learning is used to solve classification as well as regression problems. It has been mostly used as classifier in the suicide prediction and prevention models of existing literature where distance from every neighbour of a data point is measured and the number of nearest data points is denoted by a variable k based on the assumption that similar data points reside close to each other [62]. The decision of classifying the new data point is taken according to the minimum distance it has with its neighbours. Figure 4.8 shows the working mechanism of the said process.

4.3.2.4 Decision Tree Algorithms

Decision trees are constructed by an algorithm which splits a dataset based on some conditions [63]. The working mechanism is shown in Fig. 4.9. Here in suicide prediction approaches, decision tree has been applied in the questionnaires to find the subject's mental condition.

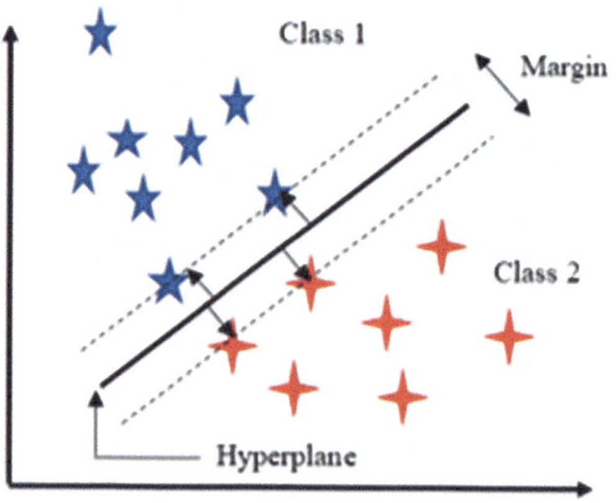

Fig. 4.7 Working mechanism of support vector machine

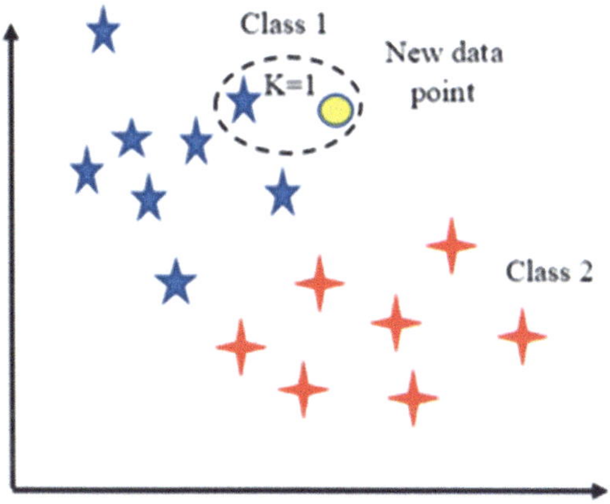

Fig. 4.8 Working mechanism of k-nearest neighbour

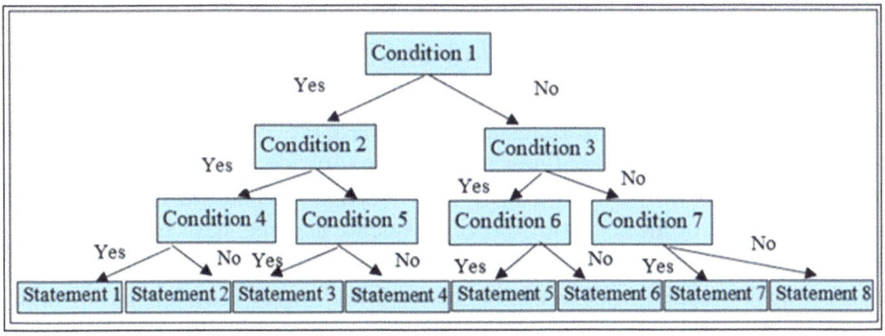

Fig. 4.9 Working mechanism of decision tree

4.3.3 Feature Selection Algorithms

Dimension reduction of features is one of the core concepts in machine learning which is performed sometimes based on filtering technique and sometimes based on wrapper technique [25] to select the most contributing features to our predicted variables or outputs. Irrelevant features may be a reason behind decrease in the accuracy of any model, and so to optimize the cost and performance of any system, feature selection is used.

4.3.4 Principal Component Analysis (PCA)

The most popular feature selection algorithm which has been widely used in suicide prediction and prevention approaches using machine learning is principal component analysis, where reduction of dimension of the feature space is done by one of the two methods, i.e. feature elimination and feature extraction [10].

The study shows that existing researchers have used supervised learning like SVM [4, 5, 10, 29], regression [26] and decision tree or unsupervised ML algorithms like neural network, deep learning algorithms or k-nearest neighbour algorithm [25–28, 35, 36].

4.4 Application Areas for Suicide Prediction

The main application domains where the suicide prediction and prevention research works have been carried out include different judgement analyses using questionnaire, analysis of different social media posts, different medical interviews and analysis of suicide notes mentioning the reasons of suicides [53, 57].

4.4.1 Suicidal Tendency Prediction Using Questionnaire

Although various machine learning approaches have proved their efficiency in performance and result, still classifying and predicting suicide attempts is a crucial task because the main issue is to get the real-time data and authentic data from different sources. In most of the scenarios, previous suicide attempts have been taken into account for predicting future suicide attempt, so for tracking previous suicide attempts and ideations, questionnaires have been designed [46] in several recent works. Two questionnaires have been used in this regard. One is Beck Hopelessness Scale or BHS and the other is Revised Suicidal Behaviors Questionnaire or SBS-R [45], to acquire knowledge of suicide risks among college students, and experimental results show that lonely students who have experienced some kind of sexual assaults somewhere are having maximum amount of suicidal thoughts and risks. Childhood abuse has also been a serious reason behind some suicide attempts shown in [47].

4.4.2 Suicidal Tendency Prediction Using Social Media Posts

The use of social media is increasing per day and people use it to share and express their feelings and opinion most of the times nowadays. Suicidal tendency has been analysed using different machine learning algorithms from different social media posts in several recent works. Three different neural networks have been used in weight optimization, namely, stochastic gradient descent, limited memory BFGS and an extended version of stochastic incline lineage in [44], and SVM has been applied as classifier. The suicidal keywords have been identified first, relevant tweets from Twitter have been searched and extracted and after that, the random data like text and images are converted to numerical features using two methods, namely, CV and Tf-idf. But due to the limitation of actual suicide notes on social media, the dataset size is very small here and language considered here is only English [54–56].

4.4.3 Suicidal Tendency Prediction Using Clinical Data and Suicide Note

A study [48] shows that 85% of patients having suicide ideations seek medical treatment before attempting suicides. So, suicide attempts may be perceived from medical records as well. US veterans are more prone to suicide than general people in the USA. To analyse this trend, machine learning methods have been applied to clinical data of that particular community in [49], and for predicting suicide ideations in the US army people, a similar research was conducted on their psychiatric data by using statistical tools in [50]. By analysing texts and word categories in clinical notes, some negative words like 'anxiety', etc., are identified as suicidal words, and not only the words on verbal or non-verbal notes but also auditory parameters like repeated pause between words, stammering and other speech-related parameters can be considered for suicide attempt analysis as mentioned by Kessler et al. [51].

4.5 Conclusion and Future Scope

This chapter demonstrates the power of technological innovations to identify a negative state of mind and for detection as well as prevention of self-destructing activities performed by a person. Smart lifestyle monitoring should be encouraged specially in a smart city environment. Prior detection of unfortunate activities like suicide may save a life, and recent technological advancement may work as blessings to pursue this process. The main emphasis of the work was on predictive analysis of occurrences of suicide with the help of new age technologies and efficient algorithms, which may be helpful for saving a costly life with prior detection. The chapter mainly focuses on depressive disorders being a prime reason of suicide. Observing the increasing activities of present youth or middle-aged persons in several social media platforms, the work may be further extended to analyse the depressive disorders from our social media context.

References

1. https://www.kaggle.com/russellyates88/suicide-rates-overview-1985-to-2016. Accessed on 31 Mar 2020, 12:00 pm, Kolkata, India.
2. https://www.who.int/health-topics/depression#tab=tab_1. Accessed on 29 Mar 2020, 10:00 pm, Kolkata, India.

3. Kamalja, K. K., & Khangar, N. V. (2017). A statistical study of suicidal behaviour of Indians. *Egyptian Journal of Forensic Sciences, 7*(1), 12.
4. Ramalingam, D., Sharma, V., & Zar, P. (2019). Study of depression analysis using machine learning techniques. *International Journal of Innovative Technology and Exploring Engineering, 8*(7C2), 187–191.
5. Liu, X., Liu, X., Sun, J., Yu, N. X., Sun, B., Li, Q., & Zhu, T. (2019). Proactive Suicide Prevention Online (PSPO): Machine identification and crisis management for Chinese social media users with suicidal thoughts and behaviours. *Journal of Medical Internet Research, 21*(5), e11705.
6. Colic, S., Richardson, D. J., Reilly, P. J., & Hasey, M. G. (2018, July). Using machine learning algorithms to enhance the management of suicide ideation. In *2018 40th Annual International Conference of the IEEE Engineering in Medicine and Biology Society* (*EMBC*) (pp. 4936–4939). IEEE.
7. Ryu, S., Lee, H., Lee, D. K., & Park, K. (2018). Use of a machine learning algorithm to predict individuals with suicide ideation in the general population. *Psychiatry Investigation, 15*(11), 1030.
8. Tadesse, M. M., Lin, H., Xu, B., & Liang, Y. (2020). MDPI. *Algorithms, 13*, 7. https://doi.org/10.3390/a13010007.
9. Ryu, S., Lee, H., Lee, D. K., Kim, S. W., & Kim, C. E. (2019). Detection of suicide attempters among suicide ideators using machine learning. *Psychiatry Investigation, 16*(8), 588.
10. Iliou, T., Konstantopoulou, G., Lymperopoulou, C., Anastasopoulos, K., Anastassopoulos, G., Margounakis, D., & Lymberopoulos, D. (2019, May). Machine learning data preprocessing method for suicide prediction from family history. In *IFIP International conference on artificial intelligence applications and innovations* (pp. 512–519). Cham: Springer.
11. Walsh, C. G., Ribeiro, J. D., & Franklin, J. C. (2017). Predicting risk of suicide attempts over time through machine learning. *Clinical Psychological Science, 5*(3), 457–469.
12. Walsh, C. G., Ribeiro, J. D., & Franklin, J. C. (2018). Predicting suicide attempts in adolescents with longitudinal clinical data and machine learning. *Journal of Child Psychology and Psychiatry, 59*(12), 1261–1270.
13. Jung, J. S., Park, S. J., Kim, E. Y., Na, K. S., Kim, Y. J., & Kim, K. G. (2019). Prediction models for high risk of suicide in Korean adolescents using machine learning techniques. *PLoS One, 14*(6), e0217639.
14. Gradus, J. L., Rosellini, A. J., Horváth-Puhó, E., Street, A. E., Galatzer-Levy, I., Jiang, T., & Sørensen, H. T. (2020). Prediction of sex-specific suicide risk using machine learning and single-payer health care registry data from Denmark. *JAMA Psychiatry, 77*(1), 25–34.
15. Amin, I., & Syed, S. (2017). Prediction of suicide causes in India using machine learning. *Journal of Independent Studies and Research* (*JISR*), *15*(2), 1–7.
16. Ji, S., Pan, S., Li, X., Cambria, E., Long, G., & Huang, Z. (2019). Suicidal ideation detection: A review of machine learning methods and applications. arXiv preprint arXiv:1910.12611.
17. Rakesh, G. (2017). Suicide prediction with machine learning. *American Journal of Psychiatry Residents' Journal, 12*(1), 15–17.
18. Ji, S., Yu, C. P., Fung, S. F., Pan, S., & Long, G. (2018). Supervised learning for suicidal ideation detection in online user content. *Complexity*, 2018.
19. Varathan, K. D., & Talib, N. (2014, August). Suicide detection system based on Twitter. In *2014 Science and information conference* (pp. 785–788). IEEE.
20. Liakata, M., Kim, J. H., Saha, S., Hastings, J., & Rebholz- Schuhmann, D. (2012). Three hybrid classifiers for the detection of emotions in suicide notes. *Biomedical Informatics Insights*, *5*, BII-S8967.
21. Braithwaite, S. R., Giraud-Carrier, C., West, J., Barnes, M. D., & Hanson, C. L. (2016). Validating machine learning algorithms for Twitter data against established measures of suicidality. *JMIR Mental Health, 3*(2), e21.

22. Coppersmith, G., Leary, R., Crutchley, P., & Fine, A. (2018). Natural language processing of social media as screening for suicide risk. *Biomedical Informatics Insights, 10*, 1178222618792860.
23. Sawhney, R., Manchanda, P., Mathur, P., Shah, R., & Singh, R. (2018, October). Exploring and learning suicidal ideation connotations on social media with deep learning. In *Proceedings of the 9th Workshop on computational approaches to subjectivity, sentiment and social media analysis* (pp. 167–175).
24. Morales, M., Dey, P., Theisen, T., Belitz, D., & Chernova, N. (2019, June). An investigation of deep learning systems for suicide risk assessment. In *Proceedings of the Sixth workshop on computational linguistics and clinical psychology* (pp. 177–181).
25. Tran, T., Phung, D., Luo, W., Harvey, R., Berk, M., & Venkatesh, S. (2013, August). An integrated framework for suicide risk prediction. In *Proceedings of the 19th ACM SIGKDD international conference on Knowledge discovery and data mining* (pp. 1410–1418).
26. Nguyen, T., Tran, T., Gopakumar, S., Phung, D., & Venkatesh, S. (2016). An evaluation of randomized machine learning methods for redundant data: Predicting short and medium-term suicide risk from administrative records and risk assessments. arXiv preprint arXiv:1605.01116.
27. Berrouiguet, S., Billot, R., Lenca, P., Tanguy, P., Baca-Garcia, E., Simonnet, M., & Gourvennec, B. (2016, June). Toward e-health applications for suicide prevention. In *2016 IEEE First international conference on connected health: Applications, systems and engineering technologies (CHASE)* (pp. 346–347). IEEE.
28. Bhat, H. S., & Goldman-Mellor, S. J. (2017). Predicting adolescent suicide attempts with neural networks. arXiv preprint arXiv:1711.10057.
29. Birjali, M., Beni-Hssane, A., & Erritali, M. (2017). Machine learning and semantic sentiment analysis based algorithms for suicide sentiment prediction in social networks. *Procedia Computer Science, 113*, 65–72.
30. Fonseka, T. M., Bhat, V., & Kennedy, S. H. (2019). The utility of artificial intelligence in suicide risk prediction and the management of suicidal behaviours. *Australian and New Zealand Journal of Psychiatry, 53*(10), 954–964.
31. "Major depression," NAMH, p. 1, 2019.
32. Emmelkamp, P. M. (2005). Technological innovations in clinical assessment and psychotherapy. *Psychotherapy and Psychosomatics, 74*(6), 336–343.
33. Caldwell, T. M., Jorm, A. F., & Dear, K. B. (2004). Suicide and mental health in rural, remote and metropolitan areas in Australia. *Medical Journal of Australia, 181*, S10–S14.
34. Parslow, R. A., & Jorm, A. F. (2000). Who uses mental health services in Australia? An analysis of data from the National Survey of Mental Health and Wellbeing. *Australian and New Zealand Journal of Psychiatry, 34*(6), 997–1008.
35. Wang, P. S., Aguilar-Gaxiola, S., Alonso, J., Angermeyer, M. C., Borges, G., Bromet, E. J., Bruffaerts, R., de Girolamo, G., de Graaf, R., Gureje, O., Haro, J. M., Karam, E. G., Kessler, R. C., Kovess, V., Lane, M. C., Lee, S., Levinson, D., Ono, Y., Petukhova, M., Posada-Villa, J., Seedat, S., & Wells, J. E. (2007). Use of mental health services for anxiety, mood, and substance disorders in 17 countries in the WHO world mental health surveys. *The Lancet, 370*(9590), 841–850.
36. Huang, Y. P., Goh, T., & Liew, C. L. (2007, December). Hunting suicide notes in web 2.0-preliminary findings. In *Ninth IEEE International Symposium on Multimedia Workshops (ISMW 2007)* (pp. 517–521). IEEE.
37. Marques, A., Peralta, M., Henriques-Neto, D., Frasquilho, D., Gouveira, É. R., & Gomez-Baya, D. (2020, February 6). Active commuting and depression symptoms in adults: A systematic review. *International Journal of Environmental Research and Public Health, 17*(3), 1041.
38. Cagliostro, D. (2020, January). Persistent sadness & loss of interest in life. https://www.psycom.net/depression.central.html. Accessed on 16 Apr 2020, 11:05 pm, Kolkata, India.
39. Cuijpers, P., Stringaris, A., & Wolpert, M. (2020, February). Treatment outcomes for depression: Challenges and opportunities. *Lancet Psychiatry, 2020*, 1–2. https://doi.org/10.1016/S2215-0366(20)30036-5.

40. Lindberg, L., et al. (2020, March). Anxiety and depression in children and adolescents with obesity: A nationwide study in Sweden. *BMC Medicine, 18*, 30. https://doi.org/10.1186/s12916-020-1498-z.
41. Islam R, Kabir A,Wang H, Ulhaq A (2019), Depression detection from social network data using machine learningTechniques, Islam et al. Health Inf Sci Syst, 6(8), pp. 1–12
42. Marks, M. (2019, January 29). Artificial intelligence based suicide prediction. *18 Yale Journal of Health Policy, Law, and Ethics, 98*, 98–121.
43. Marques, A., Peralta, M., Henriques-Neto, D., Frasquilho, D., Rubio Gouveira, É., & Gomez-Baya, D. (2020). Active commuting and depression symptoms in adults: A systematic review. *International Journal of Environmental Research and Public Health, 17*, 1041. https://doi.org/10.3390/ijerph17031041, 1–11.
44. Shahreen, N., Subhani, M., & Rahman, M. M. (2018, September). Suicidal trend analysis of twitter using machine learning and neural network. In *2018 International Conference on Bangla Speech and Language Processing* (*ICBSLP*) (pp. 1–5). IEEE.
45. Korrapati, R., Nuthalapati, K., & Thenmalar, S. (2018). A survey paper on suicide analysis. *International Journal of Pure and Applied Mathematics, 118*(22), 239–244.
46. Oh, J., Yun, K., Hwang, J. H., & Chae, J. H. (2017). Classification of suicide attempts through a machine learning algorithm based on multiple systemic psychiatric scales. *Frontiers in Psychiatry, 8*, 192.
47. Sachs-Ericsson, N. J., Stanley, I. H., Sheffler, J. L., Selby, E., & Joiner, T. E. (2017). Non-violent and violent forms of childhood abuse in the prediction of suicide attempts: Direct or indirect effects through psychiatric disorders. *Journal of Affective Disorders, 215*, 15–22.
48. Ostacher, M. J., et al. (2015). A clinical measure of suicidal ideation, suicidal behavior, and associated symptoms in bipolar disorder: Psychometric properties of the Concise Health Risk Tracking Self-Report (CHRT-SR). *Journal of Psychiatric Research, 71*, 126–133.
49. Ben-Ari, A., & Hammond, K. (2015). Text mining the EMR for modeling and predicting suicidal behavior among US Veterans of the 1991 Persian Gulf War. In *2015 48th Hawaii International Conference on System Sciences* (*HICSS*) (pp. 3168–3175). Kauai, HI, USA. https://doi.org/10.1109/HICSS.2015.382.
50. Kessler, R. C., et al. (2015). Predicting U.S. Army suicides after hospitalizations with psychiatric diagnoses in the Army Study to Assess Risk and Resilience in Service members (Army STARRS). *JAMA Psychiatry, 72*(1), 49–57. https://doi.org/10.1001/jamapsychiatry.2014.1754.
51. Kessler, R. C., Warner, L. C. H., Ivany, L. C., Petukhova, M. V., Rose, S., Bromet, E. J., Brown, L. M., III, Cai, T., Colpe, L. J., Cox, K. L., & Fullerton, C. S. (2015). Predicting US Army suicides after hospitalizations with psychiatric diagnoses in the Army Study to Assess Risk and Resilience in Service members (Army STARRS). *JAMA Psychiatry, 72*(1), 49.
52. Amin, I., & Syed, S. (2017). Prediction of suicide causes in India using machine learning. *Journal of Independent Studies and Research – Computing, 15*(2), 1–6.
53. Burke, T. A., et al. (2020). Using machine learning to classify suicide attempt history among youth in medical care settings. *Journal of Affective Disorders, 268*(1), 206–214. https://doi.org/10.1016/j.jad.2020.02.048.
54. Iliou, T., et al. (2016). Machine learning preprocessing method for suicide prediction. In L. Iliadis & I. Maglogiannis (Eds.), *Artificial intelligence applications and innovations. AIAI 2016. IFIP Advances in information and communication technology* (Vol. 475). Cham: Springer. https://doi.org/10.1007/978-3-319-44944-9_5.
55. Su, C., Aseltine, R., Doshi, R. et al. Machine learning for suicide risk prediction in children and adolescents with electronic health records. Transl Psychiatry 10, 413 (2020) pp.1-10. https://doi.org/10.1038/s41398-020-01100-0
56. Walsh, C. G., Ribeiro, J. D., & Franklin, J. C. (2017). Predicting risk of suicide attempts over time through machine learning. *Clinical Psychological Science, 5*(3), 216770261769156.
57. Bradvik, L. (2018). Suicide risk and mental disorders. *International Journal of Environmental Research and Public Health, 15*(9), 2028.

58. Torous, J., et al. (2018). Smartphones, sensors, and machine learning to advance real-time prediction and interventions for suicide prevention: A review of current progress and next steps. *Current Psychiatry Reports, 20*(7), 51. https://doi.org/10.1007/s11920-018-0914-y.
59. Kleiman, E. M., Turner, B. J., Fedor, S., Beale, E. E., Huffman, J. C., & Nock, M. K. (2017). Examination of real-time fluctuations in suicidal ideation and its risk factors: Results from two ecological momentary assessment studies. *Journal of Abnormal Psychology, 126*(6), 726.
60. Berrouiguet, S., Larsen, M. E., Mesmeur, C., Gravey, M., Billot, R., Walter, M., et al. (2018). Toward mHealth brief contact interventions in suicide prevention: Case series from the Suicide Intervention Assisted by Messages (SIAM) randomized controlled trial. *JMIR mHealth and uHealth, 6*(1), e8.
61. Agrawal, R. (2020). Fundamentals of machine learning. In *Machine learning for healthcare: Handling and managing data* (p. 1).
62. Agrawal, R. (2019). Integrated parallel k-nearest neighbor algorithm. In *Smart intelligent computing and applications* (pp. 479–486). Singapore: Springer.
63. Batra, M., & Agrawal, R. (2018). Comparative analysis of decision tree algorithms. In *Nature inspired computing* (pp. 31–36). Singapore: Springer.

Chapter 5
A Review of Particle Swarm Optimization in Cloud Computing

Devaraj Verma C, Harshvardhan Tiwari, and Madhumala RB

5.1 Introduction

Leasing the computational resources on demand is nothing new. It started in the late 1990s and is growing exponentially year by year. The resource allocation problem is an NP-hard problem, and the time taken for allocating the resources plays a vital role in defining efficiency. Linear methods are good in allocating the resources, but the time taken to allocate will increase exponentially as the number of clients requesting for the resources increase. Though the artificial intelligence and machine learning algorithms are the best fit, they suffer from the requirement that we need to have a huge amount of computational power and internal memory to efficiently allocate the resources, and hence they are economically not a feasible solution. They are even time efficient and require less computational power and internal memory. The only disadvantage is they may converge at a local optimum, i.e., the solution may not be the best. Here in this chapter, we discuss particle swarm intelligence and its applications.

Cloud computing is a technology where the cloud user is able to get the demanded cloud resources over the Internet, used to minimize the cost of computing [1]. Cloud computing is an on-demand technology for proving the quality service to the end users. Cloud computing is one of the popular options for people and business for number of reasons such as cost savings, resource management, increased productivity, increased in energy savings, and also in speed. Cloud computing provides services by which we can access the customized applications over the Internet and allows users to learn how to configure applications. Cloud users can access resources

D. V. C · M. RB (✉)
Department of Computer Science Engineering, Jain University, Bangalore, India
e-mail: rb.madhumala@jainuniversity.ac.in

H. Tiwari
CIIRC, Jyothy Institute of Technology, Bangalore, India

M. Moh et al. (eds.), *Smart IoT for Research and Industry*, EAI/Springer Innovations in Communication and Computing,
https://doi.org/10.1007/978-3-030-71485-7_5

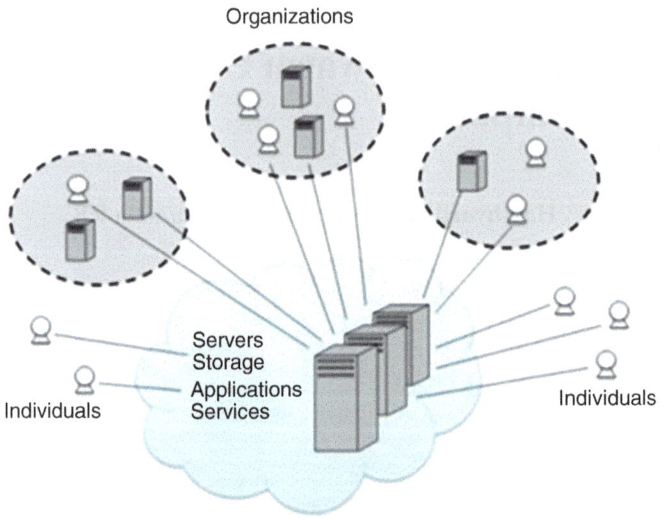

Fig. 5.1 Cloud computing architecture

through the Internet from anywhere anytime, as long as they need resources without worrying any maintenance of resources. The cloud computing refers to a network of nodes that provide services to end users on demand over the Internet. In other words, cloud computing means providing the resources at remote location. Cloud can provide services on WAN, LAN, or VPN.

Users customized applications such as e-mail, video conferencing, and customer relationship management (CRM) database applications. Cloud computing manipulates, configures, and accesses the applications online. As shown in Fig. 5.1, cloud computing offers on-demand resource infrastructure platform for its users.

5.1.1 Cloud Computing

Cloud technology is a combination of both hardware and software and database applications; users need not worry about the maintenance of the resources because it is completely taken cared of by the cloud providers,; users can just apply their technologies and they can work based on the demanded resources. A huge number of applications are available over cloud.

Many service providers are providing maximum resources over cloud with QoS policies. Security concerns are very high for all customer relationship management applications, irrespective of the domain; it may be research domain, medical domain, agricultural, military applications, and many more. Due to these

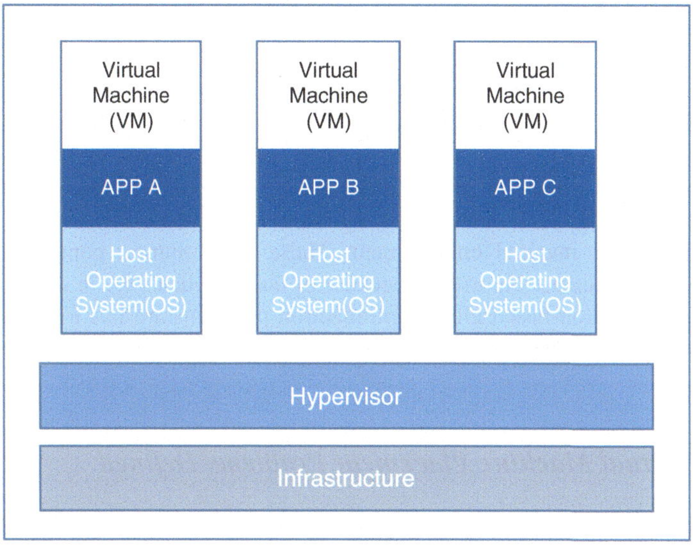

Fig. 5.2 Virtualization architecture

tremendous applications over the cloud data center, optimization of the resources over the cloud data center is a great challenge; one such key technology is the cloud virtualization technology, where virtual machines are mapped to the physical machines.

Many optimization techniques are proposed by many researchers, based on the different methodologies, few used heuristic, and few used meta-heuristic where the algorithms must be in a position to solve both single and multidimensional problems. These algorithms also solve continuous and discrete mathematical problems. Figure 5.2 shows the basic architecture of virtualization over the cloud data centers. Infrastructure as a service (IaaS) is the platform where it provides infrastructure as a service to the end users based on their need. Cloud virtualization allows the user-customized applications. Cloud computing serves both software and hardware applications based on the user demands. Only few services feasible cloud computing services and accessible to end users with less amount of time, cloud working models. IaaS is the delivery of resources over the cloud to provide infrastructure based on demand scalable service of the customer. IaaS provides fundamental resources such as physical machines, virtual machines, storage, CPU, etc., usually billed depending on the usage based on the type of service model used [2]. Platform as a service (PaaS) provides the dynamic environment for applications, development and deployment tools, etc. PaaS provides needed facilities which are required to support the life cycle of delivering web-based applications over the cloud data center services via the Internet.

5.1.2 Virtual Machines

Optimization is a technique that allows a part of our day-to-day life. In the sense, it can be called as an art of choosing the best alternative solutions among an available set of options.

Optimization is needed to improve the precopy approach based on time series [3]. In the olden decades, many global optimization methods have been proposed due to the nature-inspired environment. These are common population meta-heuristics called generic algorithms because of their applicability in a wide area of problems. Population-based global optimization problems are fast tools that overcome the limitations [4].

5.1.3 Virtual Machine Placement Problems Defined

Virtual machine placement is a method of mapping virtual machines to host machines. As virtualization is a process of cloud computing, VM selection is a vital approach for improving resources, energy efficiency, and maximum utilization in cloud services. In cloud computing, virtual machine (VM) placement problem is a critical task which includes the VM migration and intended to find the customized physical machine (PM) to host the applications where there is co-relation between the patterns to find the specific group of desktop [5]. It has an effect on the overall performance, maximizing resource utilization and reducing power consumption over the data centers and reducing maintenance cost. Many VM placement techniques are proposed for VM placement in the data centers to improve the resource utilization.

Let us consider the situation. We have four servers, which contain a quad-core processor, having a capacity of executing two VMs. The system is hosting four customized applications 1–4.

1. For every server, calculate the resource requirement using the available resources needed by each server for certain period of time, for example, 5 weeks.
2. Select a target server which has the compatible software, required CPU, network, and also the storage.
3. Place the primary virtual machine on the first server. Try to place the second virtual machine on the same old server, and check whether it can satisfy resource demands or not. If not, then add the new physical machine; accommodate the VM on this newly created machine. Continue this step so that all the VMs have been placed and add a new physical machine when it is required.
4. The resulting host machines comprise the set of the consolidated servers. Finally the total number of servers required will be reduced from four to two. If only two active VMs are running on the same server, then the CPU utilization of server is calculated as the total of the CPU utilizations over the two VMs and also the same is applicable for utilization of memory resources over the cloud data centers.

For example, consider an example of VMPM pair such as the following: (i) (30%, 40%) is a pair of the CPU cycles and memory of a VM and (45%, 50%) for the second VM. Then, the resultant server accommodates the two VMs having the configuration as (65%, 80%), i.e., vector sum. To avoid CPU and memory usage from reaching 100%, we need to have an upper bound on resources for single server by considering the threshold value. With this it is possible to achieve 100% utilization of cloud resources which can reduce the performance. VM live migration technology consumes fewer amounts of CPU cycles during the migration multi-cloud database (MCDB) [6] (Fig. 5.3).

The above figure shows the different types of virtualization techniques and how best we can employ different virtualization techniques efficiently and effectively [7]. Desktop virtualization is a technology where it allows the users to simulate their services remotely or to the devices that are connected to the workstation. This contributes the separation of desktop workstation from the client applications. This increases the simplicity of the work and the productivity of the work also reduces the downtime of the workers. Server virtualization is to divide the server into unique virtual servers based on the demand of the users and provides the virtual version of the demanded resources. Each server runs on its own operating system independently. It allows the server to increase the server ability, reduce cost in the user-customized applications, and also increase the performance ratio.

Storage virtualization is the abstraction of physical resources and housed in a central server as and when required will be allocated, i.e., based on the customer demand the virtualized resources will be allocated. It creates an abstraction layer

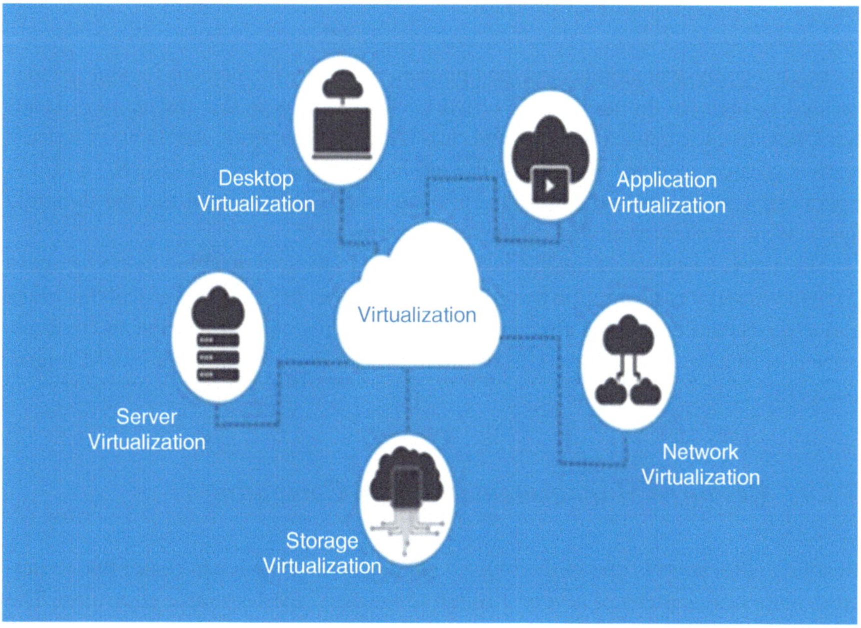

Fig. 5.3 Virtualization in cloud environment

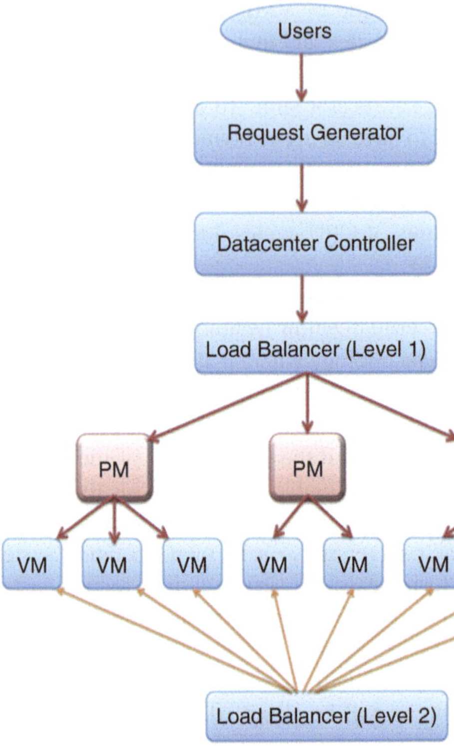

Fig. 5.4 VM placement in Cloud datacenter

between the physical storage and the operating system which means logical space is created to manage the metadata. Network virtualization is the abstraction of network resources, creates a overlay, and runs the virtual network. Application virtualization creates a virtual version of the application and interacts directly with the operating system, and with this invention it is possible to increase the dynamic application which is available on the host system.

The main objectives of the virtual machine and physical machine mapping are to (i) reduce energy consumption by reducing the number of running physical machines and to do dynamic resource allocation, (ii) improve resource utilization and to minimize cost of a data center, and (iii) improve SLA and reduce number of VM migrations (Fig. 5.4).

5.1.4 Resource Management in Cloud Computing

Resource allocation is one of the biggest challenges in cloud data center due to heterogeneity of resources; it is very difficult to manage cloud resource allocation. The above diagram shows the glimpse of the methods used in resource allocation. Primary factor here is to detect the type of resource which is in demand based on the

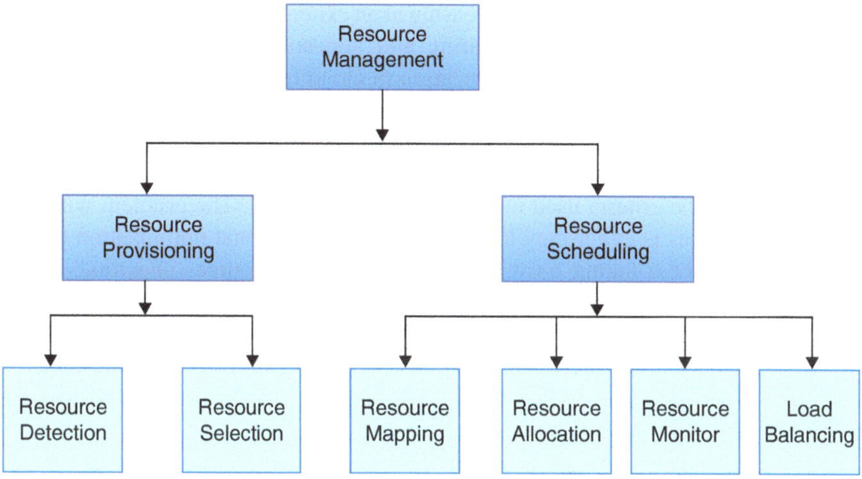

Fig. 5.5 Resource management in cloud data center

demand the resources will be allocated. Resource provisioning plays a very important role where it mainly concentrates on the quality of the data and time required to complete the task without compromising in quality of resources. Resource mapping depends on the availability of the resources and the demanded resources, and based on this ratio resource allocation will be done. Resource monitoring cloud data center is in turn providing the needed data to the cloud users depending on the cloud environment resources that will be allocated. Novel taxonomies must be proposed to balance the resources over cloud [8] (Fig. 5.5).

Many researchers explained the concept in a strategic way; the main idea here is to monitor the ratio of available and demanded resources. Once it is matched with the need, then automatically resource load balancing can be tackled. So resource monitoring and load balancing are both issues that can be solved very easily.

Cloud resource monitoring is the basic necessity before thinking about the quality of services, which mainly concentrated on the availability in all the directions. The data or the resource is not from one direction, i.e., data or information may come from different dimensions, and balancing is the major task; we need an efficient load balancing algorithm to manage the resources in an efficient manner.

5.2 Particle Swarm Optimization

Particle swarm optimization is (PSO) one of the nature-inspired population-based algorithms which utilize the swarm intelligence to find a better solution in the complete problem search space. PSO was introduced by Kennedy and Eberhart in 1995 [8], and it is a meta-heuristic powerful technique of optimization that concerns the finding of maxima or minima of functions in the possible region [32]. The number of people using cloud resources is increasing at an exponential rate; this necessitates efficient algorithms for resource sharing and allocation, and many researchers worked in this area to bring out optimization in cloud resource sharing and allocation. A tree diagram is given below to summarize the work done in this area.

Optimization is a trial and error method where classification is done based on the type of the problem, on the constraints used, on the problem size, and on the measurement using the criteria used in solving the problem [9]. The problem with the existing algorithms is either they looked at single-dimensional optimization or the optimization process taking away too much of cloud resources and time. Main objective is to study all the existing algorithms and analyze algorithm for multidimensional optimization in less amount of time and to reduce the number of running physical systems thereby increasing the power efficiency of the whole data center.

5.2.1 Parameters of PSO

To solve the said problem, we use PSO to where each particle maintains a local best and a global best solutions, and after n number of iterations, the global best solution is the selected physical machine where the virtual machine will be placed. The main objective is to place the requested VMs in such a way to reduce the number of active physical machines and the total power consumption of the data center. Being an approximation algorithm, PSO performs better when there are a lot of VM instances to be allocated on an active PM while satisfying the given objective.

The particle swarm optimization algorithm is a population-based optimization technique for solving global optimization problems, based on the social behavior. In a PSO multiple candidate solutions exist and collaborate continuously, solution is named as a particle and flies in the given problem space searching for the optimal position. A particle modifies its position as it moves from one place to another place. Its position changes according to its own experience and the experience of its neighbors. PSO is a combination of local search and the global search methods based on the particle experience in the problem search space, i.e., each iteration particle tries to update its best position. A particle status is represented by two factors: particle's position and its velocity. The new velocity and the particle's position will be updated in each iteration (Fig. 5.6).

PSO is one of the powerful optimization techniques where only few parameters are to adjust when compared to other heuristic algorithms. PSO has been applied to a wide range of applications where finding optimal solution is abundant. Datacenter has considerable amount of interest from the nature-inspired community computing that has seen too many offers which influence solving the optimization problems in multidimensional search spaces.

Fig. 5.6 Bird flocking behavior

5.2.2 *PSO Algorithm*

The PSO algorithm is having four main components which will decide the efficiency of the given algorithm, namely, initial position, velocity, weight parameters, and the fitness function. Here in this paper, we will discuss how to set the initial position and initial velocity so that the candidate solution obtained is the best one. To verify the authenticity of the arrived solution, we use the fitness function, and the fitness function ensures the PSO is optimized for the parameters we intended to.

Algorithm: Particle Swarm Optimization

```
procedure PSO
swarm ← Initialize Particles(no. of particles)
for each particle in swarm do
afterFlight(swarm)
for all itrations do
for each particle in swarm do
monitor ← Select Optimal Particle(swarm)
Post(particle, monitor)
change(particle)
afterFlight(particle)
procedure afterFlight(particle)
Check updated Particle (particle)
Compute Fitness (particle)
Update Local optimal (particle)
```

5.2.3 Modifications of the Original PSO

Particle swarm optimization has been developed by Kennedy and Eberhart in 1995 [9]. After this many researchers modified the original PSO to improve the searching capability of particles over the problem space. Many hybrid versions are introduced. Few versions are discussed here. There are many new PSO algorithms to help improve the performance of original PSO, thereby enabling the application of PSO to various optimization problems which includes constrained optimization, multi-objective optimization, and unconstrained optimization. The applications of PSO vary in complexity and cover a wide range of areas (Fig. 5.7).

The basic PSO algorithm simulates bird flocking behavior. The flight of bird flocks is simulated with good accuracy by maintaining distance between the different birds. The distance depends on size. The bird is treated as a particle and each particle is assigned a parameter called fitness value that is evaluated by a function which is optimized and has a speed corresponding to the flying of the particle.

5.3 PSO Variants

Classifications are formed depending on domain where it is applied, on attributes selected, and some other criteria. In this section, some of the different classifications of PSO algorithm along with the mathematics behind each algorithm are studied.

Fig. 5.7 Modifications of PSO

5.3.1 Continuous PSO Algorithm Techniques

Mathematics behind some of the standard PSO algorithm and its variants is summarized in Table 5.1.

5.3.2 Discrete PSO Algorithm Techniques

Initially PSO algorithm was stated as a continuous valued problem after the advancement in research and technology there came up many PSO algorithms that see it as a discrete valued problem. There are many different discrete PSO algorithms as summarized in Table 5.2.

One of the classifications of PSO is done by viewing how the particles in a group are associated, depending on some principles such as proximity, quality, and adaptability. The approaches that are mainly considered for differentiating PSO are along with its subclassification as summarized in Table 5.3.

5.3.3 PSO Analysis in Stock Market

There are many new PSO algorithms to help improve the performance of original PSO thereby enabling the application of PSO to various optimization problems which include constrained optimization, multi-objective optimization, and unconstrained optimization. The application of PSO varies in complexity and covers a wide range of areas. The basic PSO algorithm simulates bird flocking behavior. The flight of bird flocks is simulated with good accuracy by maintaining distance between the different birds. The distance depends on size. The bird is treated as a particle and each particle is assigned a parameter called fitness value that is evaluated by a function which is optimized and has a speed corresponding to the flying of the particle.

One of the applications of PSO is stock market where a huge amount of analysis is required for predicting the current stock value. In order to predict the stock values, the historic data and its relation to the market are required. This relationship is used to predict the feature stock value. The various prediction methods include the following:

- Technical analysis
- Fundamental analysis
- Traditional time series of prediction
- Machine learning methods

Technical analysis This involves predicting the appropriate time for buying or selling the stocks. The principle behind the technical analysis [15] is that share

Table 5.1 Different continuous PSO algorithm

Sl no.	Algorithm	Description	Parameters	Conclusion
1.	Standard algorithm [10]	Applies element-by-element vector multiplication	Iteration k, velocity v_k, best positions	This is basic standard algorithm used to derive specific algorithm depending on application
2.	One-dimensional algorithm [11]	Standard algorithm reduced for analysis purposes to fit for one-dimensional case	Iteration k, velocity v_k, best positions	Useful for one-dimensional case
3.	Deterministic algorithm [12]	Relationship between the random and the deterministic versions of the algorithm is established	Iteration k, velocity v_k, best positions along with attraction coefficient b	The deterministic version is obtained by setting the random numbers to their expected values
4.	Algorithm with $d = 1$ [13]	Velocity can be eliminated from standard algorithm	Iteration k, best positions along with attraction coefficient b	Objective function only depends on x
5.	Algorithm with $c = 1$ [14]	Population of particles merges the optimum location found so far	$a = 1, c = 1$	True velocity which is difference between two successive particle positions is found

Table 5.2 Different discrete PSO algorithm

Sl no.	Algorithm	Description	Parameters	Conclusion
1.	Binary particle swarm optimization [21]	Velocities are mapped to a scalar value using a sigmoidal transformation functions	Velocity	Particle positions are binary strings, while the velocities exist in continuous space
2.	Probability binary particle swarm optimization [22]	The pseudo-probability is transformed to a binary position vector, uses linear transformation	Velocity, position	Used for multidimensional knapsack problem
3.	Extended probability binary particle swarm optimization [23]	Includes mutation operator	Velocity, position, mutation	
4.	Catfish binary particle swarm optimization [24]	Considers catfish particle's position	Number of dimensions of the search space	Poor performing particles to move out of local optima for better performance
5.	Set-based particle swarm optimization [25]	Velocity as mathematical sets	Velocity, inertia weight	Generic set-based algorithm, cannot be applied to many discrete optimization problems

Table 5.3 Different PSO algorithms based on topology

Sl no.	Approach	Subtype	Description
1.	Topology [26–29]	Circle topology	Local best or ring topology 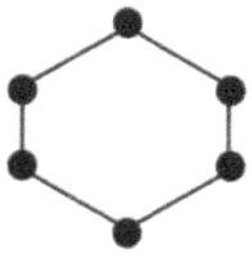
		Wheel topology	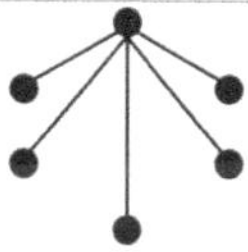
		Star topology	Also called as global best topology which is the fastest communication topology 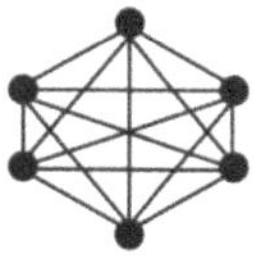
		Pyramid topology	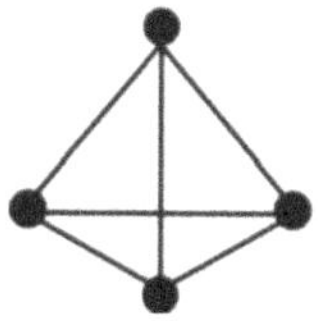
		Von Neumann topology	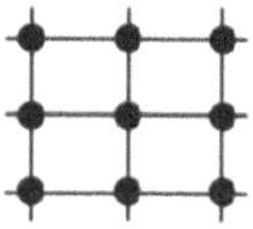
2	Social concepts [30]	Human result interaction intelligence from social	Learning from experience Adapt to the environment Determine optimal patterns of behavior and attitudes
		Culture and cognition	Mutual social learning Allows individuals to move toward adaptive patterns of behavior
3	Swarm intelligence principles [31]	Proximity principle, quality principle, diverse principle response stability, principle stability, principle adaptability	Simple space and time computations. Must respond quality factors in the environment not constrained to excessively narrow channels. Must not change when the environment changes, and when it is worth the computational price

prices change according to the trends indicated by changing attributes of investors. Technical data includes price, volume, and highest and lowest prices in the trading period which is used to predict the feature stock values. The price charts are used to detect trends which are based on supply and demand issues.

Fundamental analysis This involves applying the principles of foundation theory for selecting the individual stocks. The analyst [16] uses this method to have a clear picture of industry or market where they want to invest their wealth for gaining profit. The analysts consider parameters such as growth, dividend payout rate of interest, risk associated with investment, sales achieved, and tax rates. The main objective is to calculate the asset value. If the value of the asset is higher than the market value, then invest in it. This is helpful in predicting the market on a long-term basis.

Traditional time series of prediction This involves analyzing historical data and determines future values of a time series as a [17] linear combination. There are two types of time series forecasting:

1. Univariate traditional time series of prediction
2. Multivariate traditional time series of prediction

Which are regression models? This involves identifying a set of factors that influence the series under prediction. Univariate is based on a single variable [18], whereas multivariate depends its prediction on multiple variable [19] values.

Machine learning methods There are several methods in this category. All these methods use a set of samples for generating an approximation function [20] to generate the training data. The aim is to find conclusion from the samples in a way that when a new data is presented to the model it is possible to identify the variable that is used for representing the data.

5.4 Conclusion

The Internet of Things (IoT) involves the Internet-connected devices where we use to perform the processes and access the services. This massive information needs to be normalized. Virtualization is the key concept to balance the heterogeneous data and in turn support our way of life. Cloud computing is a technology where the cloud user can get the demanded cloud resources over the Internet. Cloud computing is an on-demand technology for proving quality service to the end users. Virtual Machine optimization is the key concept for maximum utilization of cloud resources over the cloud data center, and one such optimization algorithm is particle swarm optimization. The resource allocation problem is an NP-hard problem, and the time taken for allocating the resources plays a vital role in defining efficiency. Applying PSO for MSA belongs to the fourth sequence alignment approach. In this chapter we discussed the modifications of particle swarm optimization algorithm along with the parameter considered in each. This chapter also gives a bird's eye view on the mathematical formulae associated with each algorithm. PSO has high global convergence performance and few parameters to adjust and reduced memory and performs at an improved computation speed. These are the important reasons for PSO to be popular. The implementation of PSO is an effective solution for several problems and tons of applications in various fields including bioinformatics, and other optimization problems.

References

1. Amanatullah, Y., Lim, C., Ipung, H. P., & Juliandri, A. (2013). Toward cloud computing reference architecture: Cloud service management perspective. In *International conference on ICT for smart society* (pp. 1–4). Jakarta. https://doi.org/10.1109/ICTSS.2013.6588059.
2. Yu, W. D., Joshi, B., & Chandola, P. (2011). A service modeling approach to service requirements in SOA and cloud computing – Using a u-Healthcare system case. In *2011 IEEE 13th international conference on e-Health networking, applications and services* (pp. 233–236). Columbia. https://doi.org/10.1109/HEALTH.2011.6026754.
3. Johnson, J. A. (2013). Optimization of migration downtime of virtual machines in cloud. In *2013 Fourth international conference on computing, communications and networking technologies (ICCCNT)* (pp. 1–5). Tiruchengode. https://doi.org/10.1109/ICCCNT.2013.6726508.
4. Zhou, C., & He, G. (2008). A global convergence algorithm with stochastic search for constrained optimization problems. In *2008 Second international conference on genetic and evolutionary computing* (pp. 75–78). Hubei. https://doi.org/10.1109/WGEC.2008.40.
5. Le Thanh Man, C., & Kayashima, M. (2011). Virtual machine placement algorithm for virtualized desktop infrastructure. In *2011 IEEE international conference on cloud computing and intelligence systems* (pp. 333–337). Beijing. https://doi.org/10.1109/CCIS.2011.6045085.
6. Gupta, S., & Sharma, K. P. (2020). A review on applying tier in multi cloud database (MCDB) for security and service availability. In *2020 International conference on computer science, engineering and applications (ICCSEA)* (pp. 1–4). Gunupur, India. https://doi.org/10.1109/ICCSEA49143.2020.9132931.
7. Abdul-Rahman, O., Munetomo, M., & Akama, K. (2011). Multi-level autonomic architecture for the management of virtualized application environments in cloud platforms. In *2011 IEEE 4th international conference on cloud computing* (pp. 754–755). Washington, DC. https://doi.org/10.1109/CLOUD.2011.58.
8. López-Pires, F. (2016). Many-objective resource allocation in cloud computing datacenters. In *2016 IEEE international conference on cloud engineering workshop (IC2EW)* (pp. 213–215). Berlin. https://doi.org/10.1109/IC2EW.2016.32.
9. Kennedy, J., & Eberhart, R. (1995). Particle swarm optimization. *International Conference on Neural Networks, 1995.*
10. Madhumala, R. B., & Tiwari, H. (2020). Analysis of virtual machine placement and optimization using swarm intelligence algorithms. In A. Haldorai, A. Ramu, & S. Khan (Eds.), *Business intelligence for enterprise internet of things. EAI/Springer innovations in communication and computing*. Cham: Springer.
11. Jiang, Y., et al. (2007). An improved particle swarm optimization algorithm. *Applied Mathematics and Computation, 193*(1), 231–239.
12. Blondin, J. (2009). Particle swarm optimization: A tutorial. Retrieved from http://cs.armstrong.edu/saad/csci8100/pso tutorial.pdf
13. Clerc, M. (1999). The swarm and the queen: Towards a deterministic and adaptive particle swarm optimization. In *Proceedings of the 1999 Congress on Evolutionary Computation-CEC99 (Cat. No. 99TH8406)*.
14. Patro, S., et al. (2015). Technical analysis on financial forecasting. arXiv preprint arXiv:1503.03011.
15. Bagheri, A., Peyhani, H. M., & Akbari, M. (2014). Financial forecasting using ANFIS networks with quantum-behaved particle swarm optimization. *Expert Systems with Applications, 41*(14), 6235–6250.

16. Zhiqiang, G., Huaiqing, W., & Quan, L. (2013). Financial time series forecasting using LPP and SVM optimized by PSO. *Soft Computing, 17*(5), 805–818.
17. Hu, M., et al. (2019). Modern machine learning techniques for univariate tunnel settlement forecasting: A comparative study. *Mathematical Problems in Engineering, 2019*.
18. Han, M., et al. (2018). Multivariate chaotic time series online prediction based on improved kernel recursive least squares algorithm. *IEEE Transactions on Cybernetics, 49*(4), 1160–1172.
19. Hegazy, O., Soliman, O. S., & Salam, M. A. (2015). Comparative study between FPA, BA, MCS, ABC, and PSO algorithms in training and optimizing of LS-SVM for stock market prediction. *International Journal of Advanced Computer Research, 5*(18), 35–45.
20. Khanesar, M. A., Teshnehlab, M., & Shoorehdeli, M. A. (2007). A novel binary particle swarm optimization. In *2007 Mediterranean conference on control & automation*. IEEE Digital Library.
21. Wang, L., et al. (2008). A novel probability binary particle swarm optimization algorithm and its application. *Journal of Software, 3*(9), 28–35.
22. Liu, J., & Fan, X. (2009). The analysis and improvement of binary particle swarm optimization. In *International conference on computational intelligence and security*. IEEE
23. Chuang, L.-Y., Tsai, S.-W., & Yang, C.-H. (2011). Catfish binary particle swarm optimization for feature selection. In *International Conference on Machine Learning and Computing IPCSIT*.
24. Chou, S.-K., Jiau, M.-K., & Huang, S.-C. (2016). Stochastic set-based particle swarm optimization based on local exploration for solving the carpool service problem. *IEEE Transactions on Cybernetics, 46*(8), 1771–1783.
25. Goswami, L., Kaushik, M. K., Sikka, R., Anand, V., Prasad Sharma, K., & Singh Solanki, M. (2020). IOT based fault detection of underground cables through node MCU module. In *2020 International conference on computer science, engineering and applications (ICCSEA)* (pp. 1–6). Gunupur, India. https://doi.org/10.1109/ICCSEA49143.2020.9132893.
26. Wang, Y.-X., & Xiang, Q.-L. (2008). Particle swarms with dynamic ring topology. In *2008 IEEE Congress on Evolutionary Computation (IEEE World Congress on Computational Intelligence)*. Hindawi: IEEE.
27. Figueiredo, E. M., & Ludermir, T. B. (2012). Effect of the PSO Topologies on the Performance of the PSO-ELM. In *2012 Brazilian symposium on neural networks*. IEEE Digital Library.
28. Madhumala, R. B., Tiwari, H., & Devaraj Verma, C. (2021). A reliable frame work for virtual machine selection in cloud datacenter using particles warm optimization. *International Journal of Mathematics and Computer Science (IJMCS), 16*(2021). ISSN 1814-0432.
29. Ni, Q., & Deng, J. (2013). A new logistic dynamic particle swarm optimization algorithm based on random topology. *The Scientific World Journal, 2013*.
30. Gupta, S., Vyas, S., & Sharma, K. P. (2020). A survey on security for IoT via machine learning. In *2020 International conference on computer science, engineering and applications (ICCSEA)* (pp. 1–5). Gunupur, India. https://doi.org/10.1109/ICCSEA49143.2020.9132898.
31. Del Valle, Y., et al. (2008). Particle swarm optimization: Basic concepts, variants and applications in power systems. *IEEE Transactions on Evolutionary Computation, 12*(2), 171–195.
32. Agrawal, R. (2020). Perspectives and foundations of swarm intelligence and its application. *Swarm Intelligence Optimization: Algorithms and Applications*, 41–48.

Chapter 6
Role of Satellites in Agriculture

Prashant Johri, J. N. Singh, Sunil K. Khatri, Arko Bagchi, and E. Rajesh

6.1 Introduction

The agriculture sector in India contributes about 16% to the GDP and is the third largest contributor after the services and production sector. India is investing in the latest technologies to come up with new methods to increase the revenue of this sector.

One of the major technologies that are being widely used is the data from the satellites using multispectral and hyperspectral imaging [1]. The data from the satellites have been helping in various fields of agriculture, namely, estimating the time of harvest, anticipating in-season yields, detecting and controlling pests and diseases, understanding water and nutrient status, planning crop nutrition programmes, taking decisions about in-season irrigations, etc. The above-mentioned techniques and processes are not determined directly looking at the images and the data. An in-depth analysis must be done which involves a lot of techniques and processes.

After applying the processing techniques, we can determine the soil and crop conditions, leaf area analysis, mineral analysis, climate analysis, temperature variations, water quality, etc. Based on these results obtained, they are furthermore analysed, and trends are realised and hence decisions like irrigation requirements, fertiliser requirements, mixing of different crops to give optimum yield, the forecast for the perfect time for harvesting and cultivation, etc. These decisions, when applied to a particular area or patch of land, is called precision agriculture.

Precision agriculture uses these latest technologies driven by satellites and remote sensing, and focuses on a specific area rather than considering the whole

P. Johri (✉) · J. N. Singh · A. Bagchi · E. Rajesh
Galgotias University, Noida, UP, India
e-mail: rb.madhumala@jainuniversity.ac.in

S. K. Khatri
Amity University, Tashkant, Uzbekistan

M. Moh et al. (eds.), *Smart IoT for Research and Industry*, EAI/Springer Innovations in Communication and Computing,
https://doi.org/10.1007/978-3-030-71485-7_6

Table 6.1 Satellite types

Satellite type	Description
Open data satellites	Data from these satellites are unclassified, i.e. they are freely available for everyone to use
Commercial satellites	These satellites are used for civilian, government or nonprofit use; they are not used for military or any human space flight programme
Weather satellites	These satellites are used for collection information that is used for weather forecasting
Geodesy satellites	These satellites are used for getting information about the earth like position, measurement, variations, etc.
Ocean satellites	These satellites are used to study oceans which can be used to look over oceanic life

area since the texture, soil, humous, temperature, moisture content, etc., differ and they are not uniform throughout the same field itself; the precision agriculture uses the approach to work on those parameters area wise rather than a whole, for example, using the fertilisers only where it is required rather than spraying it all over the field.

This approach is not only helps in reducing the overall cost, but also using the inputs wherever necessary, helping in the optimised growth of the harvest and hence helping in taking decisions accurately on time. Decision plays an important role in the agricultural sector as the time to cultivate, spray pesticides and fertilisers, etc., is to be decided accurately and the untimely inputs can put all the efforts in vain. The emerging advancements that encompass data visualisation, processing and in-depth analysis aid the person on the ground on time in contrast to the decision taken solely on experience and knowledge of farmers.

These technologies bless us with great benefits, namely, cost reduction, maximising efficiency, optimum use of inputs (like seeds, insecticides, pesticides, etc.), determining the correct amount of inputs needed and hence increasing the overall efficiency. This is a huge challenge to the farmers as well as the agricultural sector administration as there have been increased changes in the weather and the climate as well as the overall changing of environments drastically (Table 6.1).

6.2 Working

The big data from the satellites is however not ready for use due to its complexity and the inability of a normal person to understand. The images from the satellites are of very large size and hence require computers with high computational and processing capabilities. Apart from the high system requirements, the images must be processed before analysing them (Fig. 6.1).

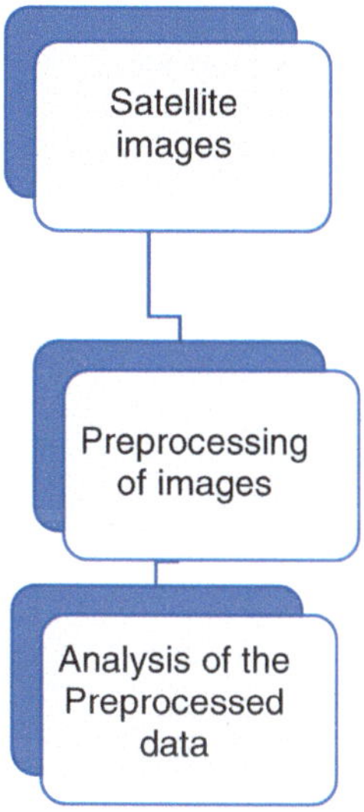

Fig. 6.1 Data processing

Every object has the ability to absorb, reflect and transmit waves; this property of the objects is exploited and used in the remote sensing technology. These interactions of objects are then analysed in the microwave, infrared and visible ZZregion.

According to Moran et al. 1997, Bastiaanssen et al. 2000, Pinter et al. 2003 and Mulla 2013, the remote sensing systems comprise visible/NIR (near-infrared = 00.4–01.5 mm) sensors for plant vegetation investigation, SWIR (short-wavelength infrared, 1.5–3 mm) sensors for plant moisture investigation, TI (thermal infrared, 3–15 mm) sensors for crop area surface or crop canopy temperature investigation and microwave sensors for soil moisture investigation.

The programmes of Moderate Resolution Imaging Spectroradiometer, also known as MODIS, National Aeronautics and Space Administration (NASA, Washington, D.C.), Landsat (NASA and United States Geological Survey, known as USGS), European satellites such as SPOT (SPOT Image, Toulouse, France), Chinese satellites (Ziyuan, China) and the Centre for Resources Satellite Data and Application (Beijing, China) all produce final outcome at distinct levels more or less relying on several applications [2] (Fig. 6.2).

6.3 Preprocessing

The data that comes is in the raw form has a lot of irregularities, and the sampling is not that consistent and contains speckles and various other anomalies. These irregularities can affect the data heavily, and the result hinders accurate results. The preprocessing includes several techniques like cleaning, integration, transformation and reduction. It is a typical name for manipulation with images at the lowest

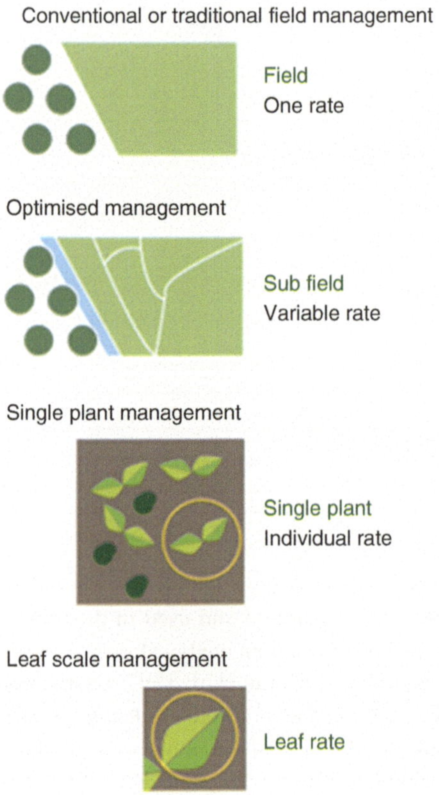

Fig. 6.2 Example of precision agriculture. (Source: Satellite Applications Catapult)

abstraction level. The purpose of preprocessing is an improvement of the image data that stops the unwanted deformations and enhances the image characteristic which are crucial for further action.

Recent studies [3, 4] revealed, how precision agriculture can be benefited from image processing. Satellite imagery data sets have become readily available by open access to NASA Landsat in 2008 [5] and to ESA Sentinel satellites.

Satellite Remote Sensing Data: Product Levels

The normalised difference vegetation Index (NDVI) is one of the attributes that is being used to process the data into a usable form to predict the concentration of chlorophyll or biomass which can later be interpreted in many ways to determine various other factors.

Brief Discussion and Review of the Analysis Techniques

There are a lot of techniques available that are being deployed for the purpose of analyzing the images. The most popular among them are deep learning, machine learning, K-means, support vector machines (SVM), artificial neural networks

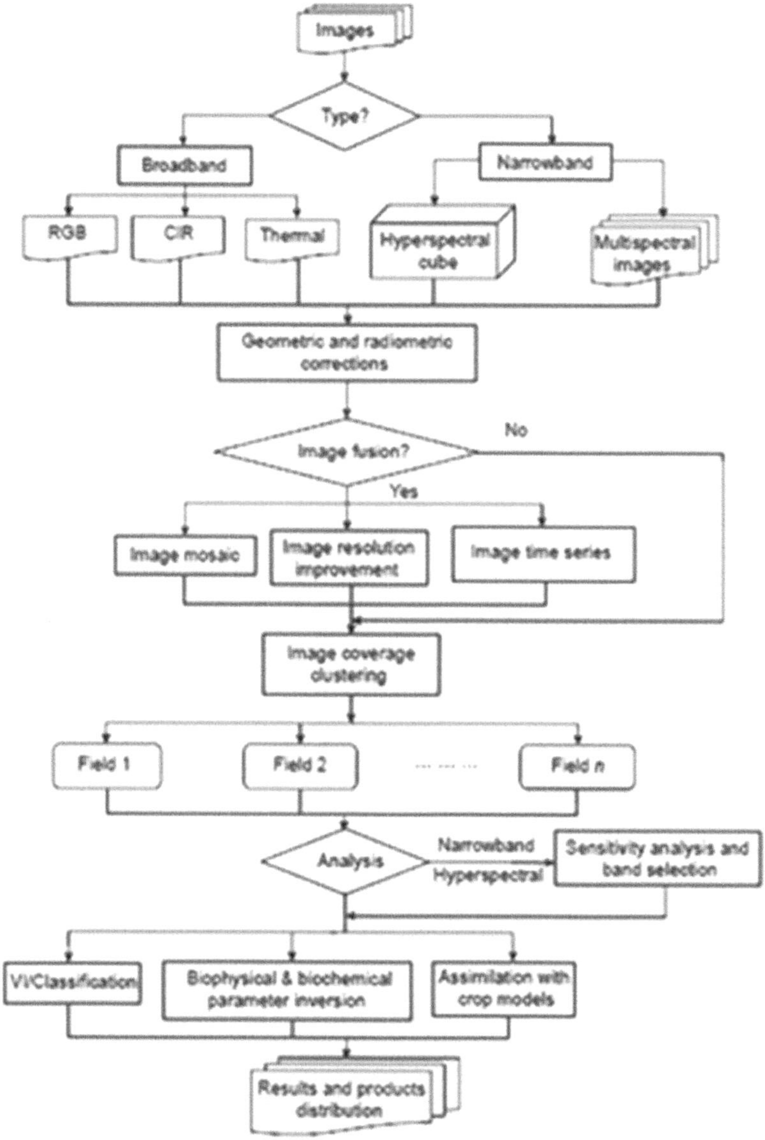

Fig. 6.3 Analysis and management flow of remote sensing [2]

(ANN), linear polarisations, wavelet-based filtering, vegetation indices (NDVI) and regression analysis [6, 7] (Fig. 6.3).

Machine learning (ML): It is a branch of science that gives the capability to learn without being rigorously programmed [8, 9]. ML models are trained using training data, and then this trained model is used to analyse the new data that has to be tested. Based on the type of learning, ML tasks can be categorised into two parts: supervised and unsupervised learning.

Supervised leaning: Here, both input and output data sets are available, and the functional relationships among the two are to be determined. This relation is achieved by using regression techniques (the most popular are the linear, logistic regression, stepwise regression). Also, Bayesian models (BM) which basically involve probability is used for accomplishing the task of supervised learning and determining the relation between the input-output.

Unsupervised learning: Here, we only have the input data set, and the underlying pattern or useful information from the data set has to be extracted. One of the most popular applications of unsupervised learning available is clustering, which enables grouping of closely related data. The technique that is most widely used to cluster data is K-means.

Other models which are used widely to analyse the learning are instance-based models (IBM), decision trees, artificial neural networks (ANN), etc. Deep ANNs or deep learning (DL) is an extremely potent area of ML and is being widely referred to in a lot of journals. It has gained its popularity in very little time and has emerged as the pioneer in the field of big data analysis (Table 6.2).

Deep learning has been one of the most popularly used techniques and has gained momentum in the past 5 years or so in the agricultural sector [11]. DL presents the data in hierarchical form, representing the convoluted deeper parts, making it easy to understand and further increasing the learning capabilities, performance and accuracy.

DL is the more in-depth version of ML (machine learning) model representing the data in hierarchical form, through several abstraction levels [11].

Deep learning has the ability to extract features from hierarchical compositions and can solve problems in less time as compared to the existing techniques. The

Table 6.2 List of algorithms

Types of algorithms	Description
Linear regression	It is used estimate real values, based on continuous variables, e.g. cost of house, upcoming taxes, etc. Eq: Yax + b
Logistic regression	It is a classification algorithm, used to assume discrete values, i.e. 0 or 1, true or false and yes or no (used to predict events which has only two results). Eq: odds = p/(1–p) = probability of event occurrence/probability of not event occurrence
Decision tree	It is used for classification problems; in this process a problem is divided into as many parts as possible, into understandable terms
Naive Bayes	It is a theorem based on prediction of the features of a particular class, which is considered unrelated to the presence of any characteristic. Eq: P(c|x) = P(X|c) P(c)/P(x)
kNN (k-nearest neighbour)	It is commonly used for classification problems; it stores all possible conditions and groups new cases according to the votes by its k neighbours [10]

Fig. 6.4 Online farm management platform that exploits computer vision and crop modelling. Fields in green are shown as being suitable for grazing and in red require further growth. (Source: Satellite Application Catapult)

edge of dl over others is because of the hierarchical composition that can be applied extensively and to various types of data sets like video audio, etc.

It has gained a lot of popularity particularly in the agricultural field due to the above-mentioned reasons and its flexibility (Fig. 6.4).

Many research papers have shown the immense ability of ML that can analyse the data as the time taken to do the tasks manually and using experience is not always reliable. The approaches of training the models can give amazing results and with accuracy and in lesser time. According to ([12], DL takes a longer time to train, but the execution and testing time is quite commendable with respect to other techniques.

Below are the few cases taken from research papers which shows its wide features where prediction related to crop was observed:

1. Automatic counting of coffee fruits on a coffee branch, using forty-two (42) colour features in a digital image illustrating coffee fruits [13].
2. Detection of cherry branches with full foliage using coloured digital images [14].
3. Identifying the young green citrus fruits under natural outdoor conditions using image and properties like coarseness, regularity and granularity [15].
4. Estimation of grassland biomass using vegetation indices, spectral bands of red and NIR [16].
5. Prediction of wheat yield within field variation using the normalised values of online predicted soil parameters and satellite NDVI [17].
6. Rice growth stage prediction and yield prediction using particular features like surface weather, soil physiochemical data with yield or development [18] (Table 6.3).

Table 6.3 Adopting methods

Types of algorithms	Description
Backpropagation	It is used for training feedforward neural networks for controlled learning; it is also responsible for solving an expression for the cost derivative function
Feedforward neural networks	These are usually fully interconnected. Each neuron in a layer is connected to every other neuron in the other layers
Convolution neural network	It is a method of combining two function by multiplying them, for mathematical reasons; it is calculating how much multiple functions correspond with each other as they pass over each other
Autoencoders	These are neural networks that are distributed directly, and they also increase the input strength at the output; they also have a layer of hidden code that explains the model
Generative adversarial networks	These are the upcoming popular ML learning model for e-commerce because of their skill to receive, understand and reconstruct visual

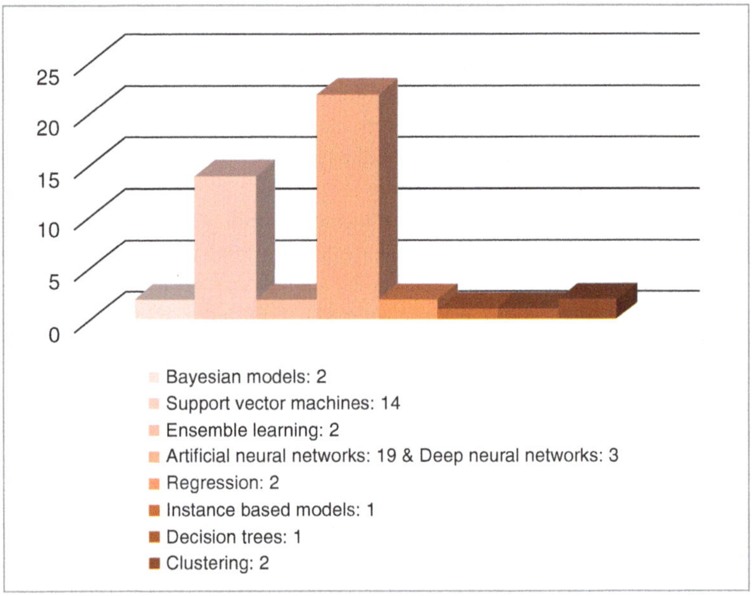

Fig. 6.5 ML models with their total rate (observed in total 40 papers) [19]

Also, diseases can also be predicted using the above-mentioned features. There are papers on weed detection too. The quality of crops has also been improved using the features stated above (Figs. 6.5 and 6.6).

The above models are largely employed into predicting the weather conditions, which further determine many other associated features with them which help immensely in the prediction of a lot of processes and decisions in agriculture. Here are a few cases:

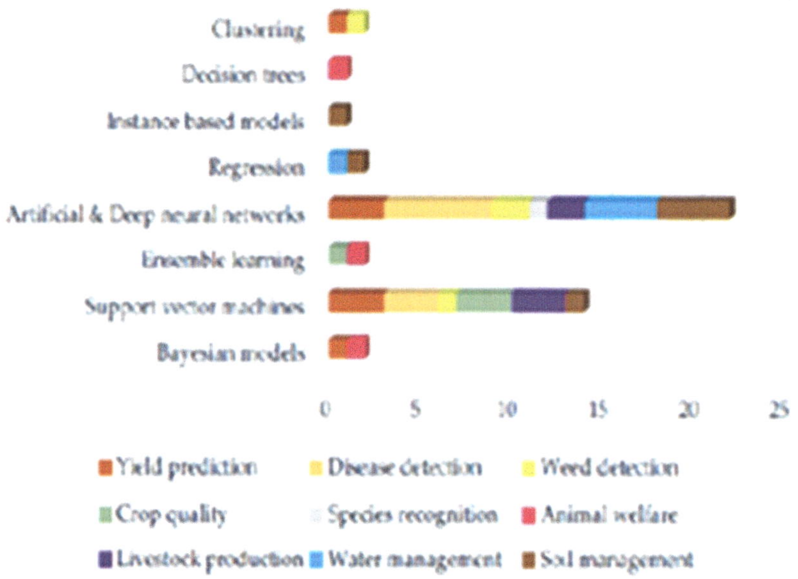

Fig. 6.6 Total number of ML models in each subcategory of four main categories [19]

1. Determining the soil texture, moisture content, soil temperature, etc., using technologies such as ANN, SVM, IBM, KNN, etc.
2. Prediction of water bodies using satellite data (the results are more pronounced in the case of large water bodies) and monitoring the quality of water using regression techniques, scenario ANN, ELM/GRNN, MARS, etc.
3. Air content monitoring, dust monitoring using data from MODIS, Sentinel, etc.

6.4 Future Scope

There are a number of challenges posed, and we seamlessly flow across into the new emerging technologies.

The data has to be reliable enough to make sure the efforts and capital do not go in vain. This can only be ensured if the data from certified and government-regulated satellites are used and are made available readily. The agricultural-related problems like seed identification, soil and leaf nitrogen content, irrigation, water erosion assessment, pest detection, herbicide use, identification of contaminants, diseases or defects on food, crop-hail damage, and greenhouse monitoring is similar to the DL according to [7], as the data analysis techniques applied on these too. Further these

Table 6.4 Available resources

Name	Website	Free access
USGS earth explorer	https://www.usgs.gov/	Yes
Sentinel open access viewer	https://scihub.copernicus.eu/	No
NASA Earthdata research	https://search.earthdata.nasa.gov/search	Yes
NOAA data access viewer	https://coast.noaa.gov/dataviewer/#/	No
DigitalGlobe open data program	https://www.digitalglobe.com/ecosystem/open-data	Yes
GEO Airbus Defense	http://www.intelligence-airbusds.com/	Yes
NASA worldview	https://worldview.earthdata.nasa.gov/?v=-59.06217605894099,-16.317711207093648,36.98469894105901,32.54947629290635	No
NOAA CLASS	https://www.bou.class.noaa.gov/saa/products/welcome;jsessionid=F33C6A66F1362C0333AFCB8486702863	Yes
National Institute for space research	https://www.natureindex.com/	Yes
Bhuvan Indian geo-platform of ISRO	https://bhuvan.nrsc.gov.in/bhuvan_links.php	No
JAXA's global ALOS 3D world	https://www.eorc.jaxa.jp/ALOS/en/aw3d30/index.htm	No
VITO vision	https://www.vito-eodata.be/PDF/portal/Application.html	No
NOAA digital coast	https://coast.noaa.gov/digitalcoast/	No
Satellite land cover	https://www.isro.gov.in/	No
UNAVCO	https://www.unavco.org/	No

technologies have to be applied in the livestock management as well at a much larger scale than now.

The big data should be made available at a rapid rate, and for this purpose, proper agricultural data management organisation should come into play with the aid and cooperation of both private and government organisations. Also, the security of this data also poses a challenge. The management and organisation of data should be done at different levels like village, local, national and global levels so that there should be uniformity and consistency with the data (Table 6.4).

6.5 Comments and Conclusion

There lies immense ocean to explore in the field of how satellites have been blessing in the field of agriculture and how it has given rise to precision agriculture. And there has to be standardisation in the acquisition of the resources. There has to be an improvement in the network speeds at which data is being uploaded and downloaded. Real-time data feed should be made available, and platforms should be available with high computational capabilities that could process the big data rapidly and in less time with efficiency. Videos from the satellites should be made real-time so that the monitoring can be real-time. The communication media between the satellites and the various other sensors and systems should be efficient and optimised, and synchronisation of these will give great results. This will also lead to the competition among organisations which will eventually lead to better outcomes in the near future (Table 6.5).

Table 6.5 Data analysis based on satellite

Satellite	Description
RADARSAT	To encourage the administration of resources (farmland included), marine observation, environment checking, ice checking, disaster management and mapping in Canada and around the globe
SMAP	To map soil moisture and freeze/thaw status
SMOS	To plan ocean surface saltiness and screen soil dampness on a worldwide scale, consequently adding to a superior comprehension of the Earth's water cycle

References

1. Ishimwe, R., Abutaleb, K., & Ahmed, F. (2014). Applications of thermal imaging in agriculture—A review. *Advanced Remote Sensing, 3*(3), 128.
2. Huang, Y., et al. (2018). *Journal of Integrative Agriculture, 17*(9), 1915–1193.
3. Jeppesen, J. H., Jacobsen, R. H., Jørgensen, R. N., Halberg, A., & Toftegaard, T. S. (2017). Identification of high-variation fields based on open satellite imagery. 11th European Conference on Precision Agriculture.
4. Jeppesen, J. H., Jacobsen, R. H., Jørgensen R. N., & Toftegaard, T. S.. (2016). *Towards data-driven precision agriculture using open data and open source software*. In International conference on Agricultural Engineering.
5. Wulder, M. A., Masek, J. G., Cohen, W. B., Loveland, T. R., & Woodcock, C. E. (2012). Opening the archive: How free data has enabled the science and monitoring promise of Landsat. *Remote Sensing of Environment, 122*(Supplement C), 2–10. Landsat Legacy Special Issue.
6. Saxena, L., & Armstrong, L. (2014). *A survey of image processing techniques for agriculture*. Proceedings of Asian Federation for Information Technology in Agriculture, Australian Society of Information and Communication Technologies in Agriculture. Perth, Australia.
7. Singh, A., Ganapathysubramanian, B., Singh, A. K., & Sarkar, S. (2016). Machine learning for high-throughput stress phenotyping in plants. *Trends in Plant Science, 21*(2), 110–124.
8. Samuel, A. L. (1959). Some studies in machine learning using the game of checkers. *IBM Journal of Research and Development, 44*, 206–226.
9. Agrawal, R., Chatterjee, J.M., Kumar, A., Rathore, P.S., & Le, D.-N. (2020). Machine Learning for Healthcare: Handling and Managing Data (1st ed.). CRC Press. https://doi.org/10.1201/9780429330131
10. Agrawal, R. (2016). A modified K-nearest neighbor algorithm using feature optimization. *International Journal of Engineering and Technology, 8*(1), 28–37.
11. LeCun, Y., Bengio, Y., & Hinton, G. (2015). Deep learning. *Nature, 521*(7553), 436–444. Lee, S. H., Chan, C. S., Wilkin, P., Remagnino, P. (2015). *Deep-plant: Plant identification with convolutional neural networks*. IEEE International Conference on Image Processing (ICIP). Quebec city, Canada, pp. 452–456.
12. Christiansen, P., Nielsen, L. N., Steen, K. A., Jørgensen, R. N., & Karstoft, H. (2016). DeepAnomaly: Combining background subtraction and deep learning for detecting obstacles and anomalies in an agricultural field. *Sensors, 16*(11), 1904.
13. Ramos, P. J., Prieto, F. A., Montoya, E. C., & Oliveros, C. E. (2017). Automatic fruit count on coffee branches using computer vision. *Computers and Electronics in Agriculture, 137*, 9–22.
14. Amatya, S., Karkee, M., Gongal, A., Zhang, Q., & Whiting, M. D. (2015). Detection of cherry tree branches with full foliage in planar architecture for automated sweet-cherry harvesting. *Biosystems Engineering, 146*, 3–15.
15. Sengupta, S., & Lee, W. S. (2014). Identification and determination of the number of immature green citrus fruit in a canopy under different ambient light conditions. *Biosystems Engineering, 117*, 51–61.
16. Ali, I., Cawkwell, F., Dwyer, E., & Green, S. (2016). Modeling managed grassland biomass estimation by using multitemporal remote sensing data—A machine learning approach. *IEEE Journal of Selected Topics in Applied Earth Observations Remote Sensing, 10*, 3254–3264.
17. Pantazi, X.-E., Moshou, D., Alexandridis, T. K., Whetton, R. L., & Mouazen, A. M. (2016). Wheat yield prediction using machine learning and advanced sensing techniques. *Computers and Electronics in Agriculture, 121*, 57–65.
18. Su, Y., Xu, H., & Yan, L. (2017). Support vector machine-based open crop model (SBOCM): Case of rice production in China. *Saudi Journal of Biological Sciences, 24*, 537–547.
19. Liakos, K. G., Busato, P., Moshou, D., Pearson, S., & Bochti, D. (2018). Machine learning in agriculture: A review. *Sensors, 18*(8), 2674.

Chapter 7
IoT in Green Engineering Transformation for Smart Cities

Shaurya Gupta and Sonali Vyas

7.1 Introduction

Sustainable progress is an ideal which leads to sustainability, which is achieved with the help of green engineering. Green engineering is defined as an engineering which is developed for the environment. In terms of industry scenario, green engineering involves strategies leading to lessening of emissions, scheme which eradicates predominantly dangerous chemicals and design to minimalise usage of regular resources along with schemes to curtail energy convention. The green engineering uprising today involves everyday operations of many manufacturing, service firms. The moral worth of any organisation to practise green engineering is a viable option by which the environment is benefited as well as the organisation also. The active mission of sustainability involves green engineering, which identifies that engineering practices remain crucial for development of applied application which encompasses the concept of sustainability in day-to-day life. The association amongst sustainable enlargement, sustainability and green engineering is depicted below.

Science in addition to technology supports sustainability by the progress of sustainable materials and practices. Expertise in technology provides pioneering solutions whenever society challenges scientists and engineers to encourage human and ecological welfare for the development activities for sustainability. Grlibler [1] debates that the mission onward is to gradually release the environment from adversarial human intervention. Green engineering is an evolving aegis which engrosses engineers from varied verticals paving ways for sustainable co-existence between humans and the world. The objective hence is sustainability in addition to green engineering is the only means to realise the objective.

S. Gupta (✉) · S. Vyas
School of Computer Science, UPES, Dehradun, Uttarakhand, India

M. Moh et al. (eds.), *Smart IoT for Research and Industry*, EAI/Springer Innovations in Communication and Computing,
https://doi.org/10.1007/978-3-030-71485-7_7

7.2 Green Engineering: Driving Forces

The principal dynamic services made by corportaes to accept green engineering besides occupational framework is considered lawful and has monetary benefit also considerations. Legal apprehensions are of highest significance for every organisation. Financial concerns also stand as one of the major significance for organisations as it involves managers' responsibility to reassure bondholders or proprietors so that corporation makes an ample turnover and enables them to base their decisions around green practices on a cost-benefit scrutiny. These organisations might agree to accept green engineering plus commercial rehearses sternly on basis of lawful and fiscal dynamics, along with the aspiration to defend the regular environment. The normative model proposed by Piaget [3], Kohlberg [2] and Rest [4, 5] involves four workings: moral awareness, moral judgement, moral motivation and moral character. Piaget, Kohlberg and others [6] have stated the significant dynamics in deciding an individual's probability of acting ethically which involves people being ethically alert, taking ethical decisions and ordering ethical standards. The philosophy of ethical progress is illustrated in Fig. 7.1.

Figure 7.1 provides various stages or levels of Kohlberg's moral development model. Each stage has an age range and a description of people belonging to that age range. The model clearly shows age range of infancy, preschool, school-age, teens and adulthood and the description of morality in the specified age range.

Kohlberg's Theory

Level/Stage	Age Range	Description
I: Obedience/Punishment	Infancy	No difference between doing the right thing and avoiding punishment
I: Self-Interest	Pre-school	Interest shifts to rewards rather than punishment – effort is made to secure greatest benefit for oneself
II: Conformity and Interpersonal Accord	School-age	The "good boy/girl" level. Effort is made to secure approval and maintain friendly relations with others
II: Authority and Social Order	School-age	Orientation toward fixed rules. The purpose of morality is maintaining the social order. Interpersonal accord is expanded to include the entire society
III: Social Contract	Teens	Mutual benefit, reciprocity. Morally right and legally right are not always the same. Utilitarian rules that make life better for everyone
III: Universal Principles	Adulthood	Morality is based on principles that transcend mutual benefit.

The Psychology Notes Headquarters - http://www.PsychologyNotesHQ.com

Fig. 7.1 Kohlberg's moral development model [2]

Table 7.1 Green engineering principles [7]

Standard I	*Designers should guarantee whatever solid and energy is ingested; besides productivities must be as harmless as possible*
Standard II	*The practices to check waste than to treat it must be designed*
Standard III	*Parting besides cleansing procedures ought to be a module of project agenda*
Standard IV	*Device machineries ought to be considered to optimise build and liveliness and meet time-based competence*
Standard V	*Device machineries ought to be output drawn relatively compared when input pushed in the form of energy besides resources*
Standard VI	*Entrenched complexity should be taken into consideration whilst finalising scheme selections on reprocess, recycle or valuable dispose*
Standard VII	*Durability ought to be the final strategic objective*
Standard VIII	*The engineering design which involves the concept "one-size-fits-all" resolutions should be preferred*
Standard IX	*Multicomponent products must strive for material unification to promote disassembly and value retention (minimise material diversity)*
Standard X	*Procedures besides schemes design ought to comprise incorporation of interconnectivity amongst flow of energy plus resources*
Standard XI	*The project must include performance designing metrics for commercial purposes*
Standard XII	*Design must rely on renewable besides freely obtainable inputs during its life course*

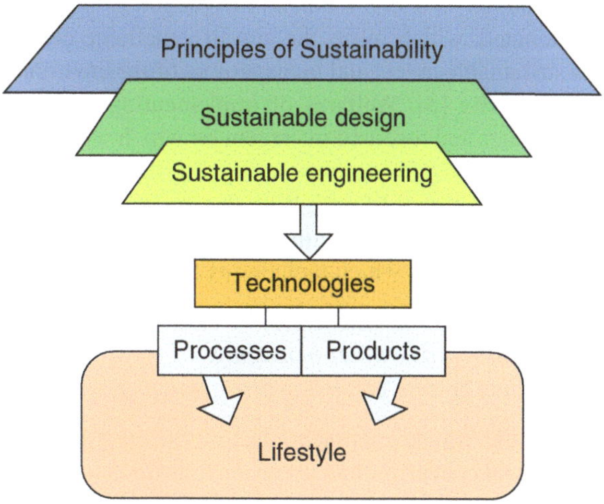

Fig. 7.2 Hierarchy of sustainability guidelines [7]

Figure 7.2 shows the principle of sustainability at the first level followed by sustainable design and engineering used in various technologies which can be depicted in processes and products. These processes and products have become an indespensable part of lifestyles of varied consumers.

- *Ethics in Green Engineering*

Green engineering involves plan and innovation besides engineering application resolutions aimed at sustainability. This method stays very extensive as it incorporates various disciplines of engineering. Ideologies in green engineering [7] deliver an outline for considerate understanding besides representing a likeness of engineering practices which makes environment supplementary sustainable. The 12 doctrines of green engineering act as a means to aid in addressing certain proposal selections which are quite applicable for essential sustainability.

Green engineering desires towards applying essentials of proposal in the following ways:

- Life-cycle contemplations.
- Multiscale applications involving procedures and schemes.

Without such an approach, a designer risks the unintended significance of doing the wrong things. In other words, one could efficiently relate principles of green engineering on a single life-cycle stage, e.g. manufacturing, etc., but still elements of sustainability would still be degraded. Therefore, life-cycle deliberations cannot be ignored. The designer of engineering clarifications must cater to various issues in addressing sustainability. The technologies have become gradually significant in talking about essential sustainability tests like energy, food, water, resource depletion, etc.

- *Internet of Things*

Within the information and communication technology field, the Internet of Things is widely debated, which fulfils the aim of cybernetic connections. It ultimately simplifies sustainable assets and management of the environment via sensor expertise plus connectivity [8]. Welfares of implementing IoT in city will offer vision of electricity water and resource management which will improve procedure performance and prognostic upkeep plus set up big data by observing systems, which promote a better quality of life [9]. With the progressive development in IoT technologies, this has been dramatically increased. The smart cities are open, efficient, connected and sustainable which includes as follows:

- *Intelligent substructure and infrastructure.*
- *Smooth flexibility and efficient water managing.*
- *Efficient supply chains besides efficient waste managing.*

Above factors are the key developments leading to the greener cities [10]. By addressing the ecological concerns of an urban design which includes technological implementation and operational risks were identified by Ejaz and Anpalagan [11], enshrined in IoT growth. IoT is continuously observed as minor enabler for GHG emissions, though the extensive acceptance lacking proper management and evaluation can have adverse environmental effects.

- *Green City Transformation with IoT*

IoT has always been an essential component of a smart metropolis. The key components in a smart town encompass information technology incorporation plus a complete use of info assets [12]. Smart city prototype is projected in Fig. 7.1 which is intended for extent valuation. This prototype entails the following:

- *Keen budget management.*
- *Smooth government domination.*
- *Keen individuals.*
- *Efficient movement.*
- *Smart existing and surroundings.*

The smart environment is the key section in the direction of a green city transformation. However, by the inbound industrialised uprising, economy is renovated to linear functioning, wherein assets are misused callously and waste is not treated adequately. Therefore, the circular economy is unified with smart and green city development and transformation which seems as a feasible resolution (Fig. 7.4).

A circular economy focusses by minimising waste and involves efficient and effective usage of natural assets and energy. It is attained via processing prototypes by tapering rounds of material and energy flows in a regenerative manner and ensuring that the ideologies of reducing and recycling are followed religiously [14]. The important activity involved here is to incorporate the following:

- *Smart commerce.*
- *Smart transportation.*
- *Smart energy flows.*
- *Smart agronomy.*
- *Smart unused assemblage in addition to leftover management.*

The circular economy has few substantial encounters [15].

- *Various IOT Roles in Dissimilar Segments*

IoT primarily involves the concept involving green smart cities. The concept of IoT in smart urban uses ICT which is aimed at worthy and quality living [16, 34]. This embraces the following considerations:

- *Improved reserve utilisation with rarer radiations.*
- *Smarter urban transportation systems improved water stock plus supply, leftover discarding amenities plus additional proficient methods for lighting and heating office block and houses.*
- *Incorporating a collaborative besides receptive urban management providing harmless community areas.*

Numerous life-threatening concerns in circular economy are linked to discarded management; Fig. 7.3 provides the role of IoT in waste management. The core issues which needs addressing are as follows:

- *Minimising the circlet to attain circuitousness.*
- *Excess retrieval and handling.*
- *Discharge justification.*

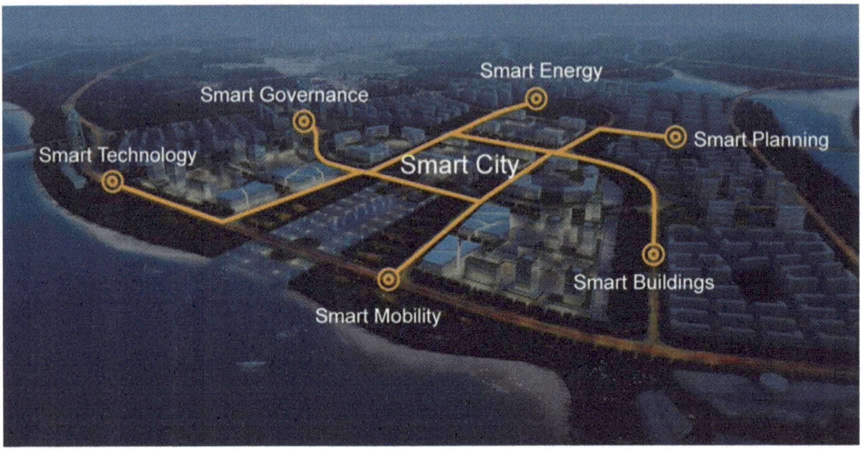

Fig. 7.3 Smart city model [13]

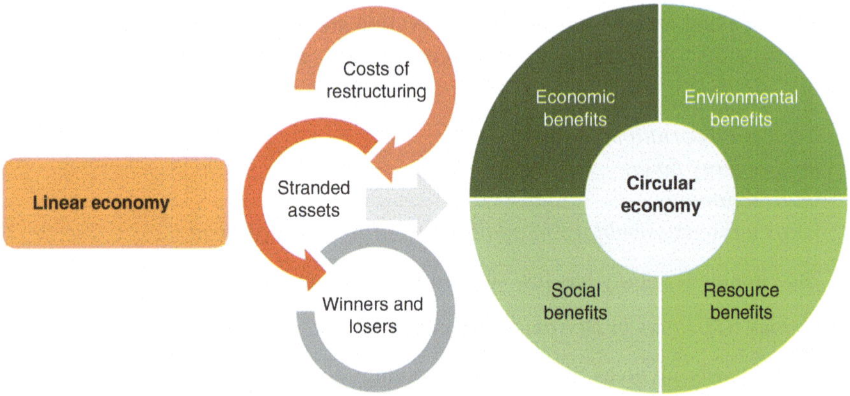

Fig. 7.4 Changeover of linear economy to circular economy [14]

The smart devices enable the leftover reduction, gathering and essential treatment and logistics depicted by Fig. 7.5. The waste management process involves the detection of waste material with the help of smart sensors. Smart sensors are vital measures of any smart structure and smart industrial units including smart transportation. Smart sensors control the illumination and aeriation of a constructed structure which is centred on temperature in addition to occupiers in a room, and smart metre keeps the energy utilisation in check and avoids needless usage of energy and in addition observation of pollution level and water leakages, and leftover recycle is also done. Therefore, operational cost is reduced and environmental issues are addressed [17]. Smart grids possess demand response mechanism to optimise dispersal of energy rendering to consumption outline and constantly monitor in

Fig. 7.5 IOT advantage in leftover managing [17]

addition to updating of information. Scheduling allocates energy wherever energy is required to maintain the system's ideal process which is attained by smart distribution system [11]. Regardless of the debated benefits of IoT implementation, there are several challenges which comprises energy ingestion and reserve abstraction besides e-waste clearance [18]. Waste heat may be used for district heating [19], and heat pumps can be collected at centres for generating cooling energy for server [20]. However, effectiveness can still be improved. The IoT has its own footprints, which have a significant impression on the environment which is perhaps considered and minimised. IoT segment desires to address energy consumption and waste generated which rises due to sector expansion [21]. The practical vision of IoT is to encounter social requirements though optimising the ecological influence. Technology competence includes safety and assets besides energy consumption which ensures sustainable application of IoT technologies [33] (Fig. 7.6).

- *Technologies for Green IoT*

Numerous technologies like green cloud computing network and RFID green tags constitute the crucial expertise for green IoT which is depicted in Fig. 7.7.

Radio frequency identification (RFID) is a minor microelectronic tool including several minor tab readers [22]. These RFID tabs store data about the items to which they are related and having broadcast range limited to metres. RFID tags are of two kinds, i.e. active and passive tabs. Active ones include batteries to constantly spread its particular indication, but the passive ones don't have battery of their own. Therefore, passive tags requires harvesting of energy from the reader signal. A

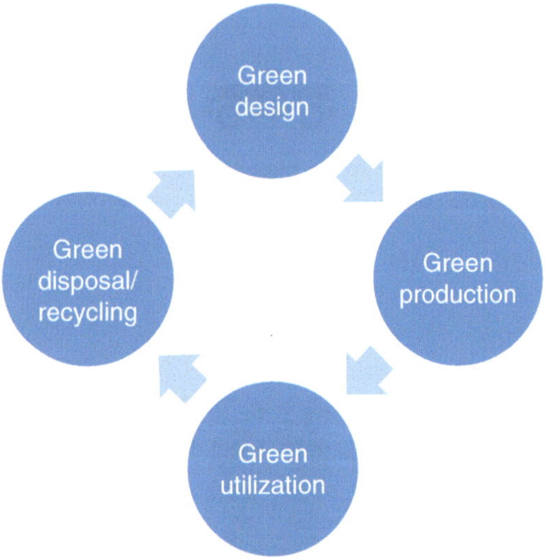

Fig. 7.6 Lifetime of green IoT

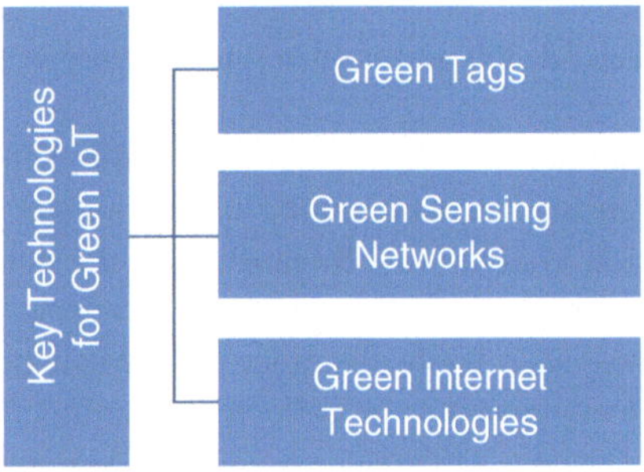

Fig. 7.7 Crucial machineries involving green IoT

projected resolution can decrease the dimension of RFID tabs and therefore minimise the amount of nondegradable material [23]. For attaining green WSN, contrasting procedures can be considered:

- *Sensor after its use should be put to snooze mode.*
- *Practice recyclable energy for device charging purposes.*
- *Practice power-efficient optimisation practices.*
- *Practice framework consciousness procedures to lessen information magnitude therefore resulting in reduction of storing capability.*
- *Practice power-efficient routing practices for lessening mobility energy ingestion.*

Manufacturers should build devices which consume a lesser amount of energy and produce greater output without reducing the performance of the devices for implementing green technology and ensuring minutest consumption of assets and thereby implementation of energy redeemable virtual technologies (Fig. 7.8).

Green IoT expertise involves an ample of amenities besides applications consisting of the following:

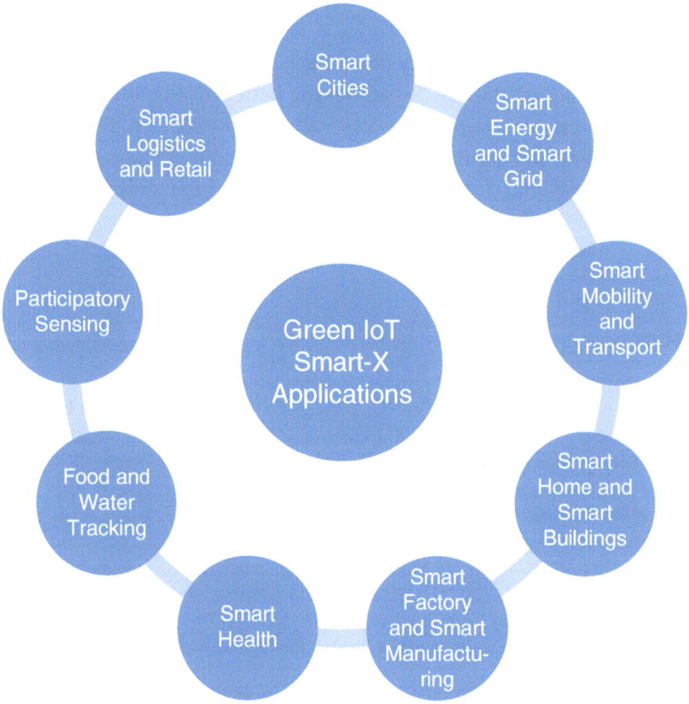

Fig. 7.8 Green IoT applications

- *Smart network schemes.*
- *Smart structures.*
- *Smart medicinal schemes.*
- *Smart transportation and logistics.*

IoT applications pertaining to smart urban involve smart parking [24, 25], environmental monitoring [26–28], traffic supervision [29] and leftover administration [30]. In [31, 32], different categorisations of IoT architecture along with challenges and infrastructure management for smart metropolises are depicted.

- *Forthcoming Research Directions and Challenges*

There are remarkable exploration ways for attaining green technology, though it's still in initial phase. Various complications and challenges which requires attention. Key challenges are as follows:

- *Integration amongst power efficiency in architecture of IoT for achieving remarkable performance.*
- *Energy consumption models must be reliable with terms and conditions of green IoT.*
- *Communicating devices should use protocols which are energy efficient with less power intake.*

- *Efficient cloud management with respect to power consumption.*

Green engineering will ultimately drive the ecosphere to sustainability, but it has to comply to laws and regulations for achieving better productivity for the organisation. Therefore, green engineering is fairly valuable for society. The actual protagonists of green engineering are the organisation's frontrunners who trust in ideologies of green engineering and follow these moralities whilst making the businesses productive.

References

1. Grlibler, A. (1998). *Technology and global change*. Cambridge: Cambridge University Press.
2. Colby, A., Kohlberg, L., Gibbs, J. T., & Lieberman, A. (1983). A longitudinal study of moral judgment. *Monographs of the Society for Research in Child Development, 200*, 46.
3. Piaget, J. (1965). *The moral judgment of the child*. New York: The Free Press.
4. Rest, J. R. (1986). *Moral development: Advances in research and theory*. New York: Praeger.
5. Rest, J. D., Narvaez, D., Bebeau, M. J., & Thoma, S. J. (1999). *Postconventional moral thinking: A Neo-Kohlbergian approach*. Mahwah: Lawrence Erlbaum Associates.
6. Duska, R., & Whelan, M. (1975). *Moral development: A guide to Piaget and Kohlberg*. New York: Paulist Press.
7. Anastas, P. T., & Zimmerman, J. B. (2003). *Environmental Science & Technology, 37*, 94A–101A.
8. Liberg, O., Sundberg, M., Wang, E., Bergman, J., & Sachs, J. (2017). *Cellular internet of things: Technologies, standards, and performance*. Cambridge, MA: Academic Press.
9. Syafrudin, M., Alfian, G., Fitriyani, N., & Rhee, J. (2018). Performance analysis of IoT-based sensor, big data processing, and machine learning model for real-time monitoring system in automotive manufacturing. *Sensors, 18*(9), 2946.
10. Conti, M., Dehghantanha, A., Franke, K., & Watson, S. (2018). Internet of things security and forensics: Challenges and opportunities. *Future Generation Computer System, 78*, 544–546.
11. Ejaz, W., & Anpalagan, A. (2019). Internet of things for smart cities: Overview and key challenges. In *Internet of things for smart cities* (pp. 1–15). Cham: Springer.
12. Kim, T. H., Ramos, C., & Mohammed, S. (2017). Smart city and IoT. *Future Generation Computer System, 76*, 159–162.
13. TU. (2015). *European smart cities 4.0*. www.smartcities.eu/index.php?cid=6&ver=4&city=94. Last accessed 28 Mar 2019.
14. Tura, N., Hanski, J., Ahola, T., Ståhle, M., Piiparinen, S., & Valkokari, P. (2019). Unlocking circular business: A framework of barriers and drivers. *Journal of Cleaner Production, 212*, 90–98.
15. Korhonen, J., Honkasalo, A., & Seppälä, J. (2018). Circular economy: The concept and its limitations. *Ecological Economics, 143*, 37–46.
16. European Commission. (2018). *Smart cities*. ec.europa.eu/digitalsingle-market/en/smart-cities. Last accessed 28 Mar 2019.
17. Garcia-Ojeda, J. C., Bertok, B., & Friedler, F. (2012). Planning evacuation routes with the P-graph framework. *Chemical Engineering Transactions, 29*, 1531–1536.
18. Wahlroos, M., Pärssinen, M., Manner, J., & Syri, S. (2017). Utilizing data center waste heat in district heating–impacts on energy efficiency and prospects for low-temperature district heating networks. *Energy, 140*, 1228–1238.
19. Biba, E. (2017). *The city where the internet warms people's homes*. www.bbc.com/future/story/20171013-where-data-centres-store-info%2D%2Dand-heat-homes. Last accessed 29 Mar 2019.

20. IBM. 2015. *IBM wants to cool data centers with their own waste heat.* www.extremetech.com/extreme/211008-ibm-wants-to-cool-datacenters-with-their-own-waste-heat. Last accessed 29 Mar 2019.
21. Klemeš, J. J., & Fan, Y. V. (2018). Internet of Things for Green City transformation. *Plenary lecture*, 25 October 2018. 4th ICLCA 2018, Johor Bahru, Malaysia. www.utm.my/iclca/iclca-2018/plenarylectures/. Last accessed 28 Mar 2019.
22. Li, T., Wu, S., Chen, S., & Yang, M. (2012). Generalized energy-efficient algorithms for the RFID estimation problem. *IEEE ACM Transactions on Networking, 20*(6), 1978–1990.
23. Shaikh, F., Zeadally, S., & Exposito, E. (2017). Enabling technologies for green Internet of Things. *IEEE Systems Journal, 11*(2), 983–994.
24. Ahlgren, B., Hidell, M., & Ngai, E. (2016). Internet of things for smart cities: Interoperability and open data. *IEEE Internet Computing, 20*(6), 52–56.
25. Ramaswamy, P. (2016). *IoT Smart Parking Systems for Reducing Green House Gas Emission.* 2016 international conference on Recent Trends in Information Technology, 2016.
26. Zhou, J., Leppnen, T., Harjula, E., Yu, C., Jin, H., & Yang, L. T. (2013). *Cloud things: A common architecture for integrating the Internet of Things with cloud computing.* Proceedings of the 2013 IEEE 17th international conference on Computer Supported Cooperative Work in Design, pp. 651–657.
27. Montgomery, B. (2015). Future shock: IoT benefits beyond traffic and lighting energy optimization. *IEEE Consumer Electronics Magazine, 4*(4), 98–100.
28. Fang, S., Xu, L., Zhu, Y., Ahati, J., Pei, H., Yan, J., & Liu, Z. (2014). An integrated system for regional environmental monitoring and management based on internet of things. *IEEE Transaction on Industrial Informatics, 10*(2), 1596–1605.
29. Mahalank, S., Malagund, K., & Banakar, R. (2016). *Device to device interaction analysis in IoT based smart traffic management system, an experimental approach.* 2016 Symposium on Colossal Data Analysis and Networking, 2016.
30. Shyam, G., Manvi, S., & Bharti, P. (2017). *Smart waste management using Internet of Things (IoT).* 2nd international conference on Computing and Communications Technologies.
31. Ganchev, I., Ji, Z., & O'Droma, M. (2014). *A generic IoT architecture for smart cities.* Irish Signals & Systems Conference 2014 and 2014 China-Ireland International Conference on Information Technologies, Ireland.
32. Sores, P., Santana, J., Sanchez, L., Lanza, J., & Munoz, L. (2017). Practical lessons from the deployment and management of a smart city Internet-of-Things infrastructure: The smart Santander testbed case. *IEEE Access, 5*, 14309–14322.
33. Murugesan, S. (2008). Harnessing green IT: Principles and practices. *IEEE IT Professional, 10*(1), 24–33.
34. Bhushan, D., & Agrawal, R. (2020). Security challenges for designing wearable and IoT solutions. In *A handbook of internet of things in biomedical and cyber physical system* (pp. 109–138). Cham: Springer.

Chapter 8
A Study on Optimal Framework with Fog Computing for Smart City

Govind Murari Upadhyay and Shashikant Gupta

8.1 Introduction

In this era of wireless communication, technology users are more dependent on the Internet and smart devices. By this the individuals and industries are producing huge volumes of data. The smart devices are producing data by their applications and sensors. These smart devices are connected through the Internet. In every sense, these devices are generating and storing large volumes of data. These end user devices connected through the internet are the Internet of Things (IoT). IoT is integrated into our daily life, for instance, healthcare system, smart cities, smart homes, industrial automation, emergency responses, and transportation. The IoT accredits things to sense and observe the environment. The data sensed by these devices and stored on servers are called clouds. The cloud system has been identified as a major provider of IoT applications as well as more storage and processing capacity. Cloud servers are physically located far away from end users. In the cloud-supported IoT system, the end user has long distance from the cloud. Due to this physical distance, there are various challenges in this IoT-cloud system like a heavy load on the cloud, high response time, and lack of global mobility. Now around 50 billion devices are online. These devices are producing unusual volumes of data because of IoT. All this data would have to be sent to the cloud server for processing and storage. In the current scenario, the cloud server cannot store huge unusual data. This huge volume of data has to be preprocessed and filtered at the middle layer before this has to be sent to the cloud for storing. Real-time data must be processed within the desired time frame because the time taken for processing and transferring data through the cloud has high latency. If this sends the entire data accumulated by the end user devices, the network traffic will be gigantic. This huge volume of data transmission

G. M. Upadhyay (✉) · S. Gupta
School of Engineering and Technology, ITM University, Gwalior, India
e-mail: shashikantgupta@itmuniversity.ac.in

M. Moh et al. (eds.), *Smart IoT for Research and Industry*, EAI/Springer Innovations in Communication and Computing,
https://doi.org/10.1007/978-3-030-71485-7_8

will consume bandwidth. For these problems, we require fog computing along with a cloud for IoT devices. Fog computing is a layer between the sensor devices and the cloud server. Sensor data is to be processed at the fog node before transmitting it to the cloud server. The time-sensitive data will be processed in the fog node. The data which is less time-sensitive would be sent to the cloud server. For the time-sensitive data, fog node will process and revert the processed data to the IoT devices and for future processing again send it to the cloud server. Fog provides transient storage before transferring to the cloud so that it can make quick decisions for time-sensitive data. Fog computing is a recent computing paradigm that is extending cloud computing towards the edge of the network. The evolution of fog computing is just to enhance the computing capabilities and improve the performance of cloud computing; it is not to replace cloud computing. The idea of introducing the fog computing concept is to add a processing and storage layer closer to the end user to improve the latency. The closeness of the fog layer near to the IoT devices will improve the quality of services, provide temporary storage of data, support high mobility, and improve latency.

8.2 Fog Architecture

To perform the implementation of fog computing, the three-layer network architecture is required. The layers are sensor layer, fog layer, and cloud layer. In the following architecture, the "things layer" have the sensors/actuators, which sense data from the environment. The sensed data is for storing or processing that can be transmitted to either the fog layer or the cloud layer. There are two types of task which have been processed, one is the light processing task, and the other is a heavy processing task. The light processing tasks (e.g., room temperature sensor) generated by the things are less time-sensitive stored data and processed by the cloud layer, and heavy processing tasks (e.g., video footage generated by a surveillance camera) are more time-sensitive, processed by fog layer. When there are light processing tasks, they have to be transferred to the cloud. These tasks can be transferred directly from the actuators to the cloud because the IoT devices are also connected to the cloud server directly or through the middle layer which is the fog layer. In case of heavy processing task processed are temporarily stored at fog layer, if a fog node is busy or overburdened through the process, then the task is to be transferred to the other fog node for processing and storing. This concept is known as task offloading. Task offloading will work on policy; when the estimated wait time is less than the threshold time, the task will be processed by the same node, and if the estimated wait time is greater than the threshold, the task is offloaded to the best neighbor. In case no fog node is free for processing, the task is transmitted to the cloud for processing and storing [1].

8.2.1 Three-Layer Architecture

Fog computing architecture is similar to other large scales of distributed architecture. This architecture comprises three layers as shown in Fig. 8.1:

(i) Things layer, (ii) fog layer, and (iii) cloud layer.

(i) Things layer: Things layer is also known as the sensor layer and obtains data through sensing using radiofrequency identification (RFID), sensors, wireless identification and sensing platform (WISP), actuators, etc. IoT nodes do not have the processing capability. The sensor node transmits the accumulated data for processing locally to the fog node or the cloud node.

(ii) Fog layer: The sensed information is sent to the second layer. Fog node processes the request by itself or sends that request to another fog node in the same domain if the waiting time is more. The fog node processed the task in the same domain/fog layer if waiting time is more in the same domain the task is forwarded to the cloud. The fog node facilitates routing in the network.

(iii) Cloud layer: The cloud layer receives the request from either fog layer or IoT nodes. This layer provides storage and processing. Processed requests can be sent back from where it was initiated [2].

Fog computing aim is to put the intelligent layer closer to the end user devices. The main focus of fog computing is to improve the quality of services (QoS) and improve latency and availability of processing nodes nearer to the sensors. The local information can be received by the end user in real time [3].

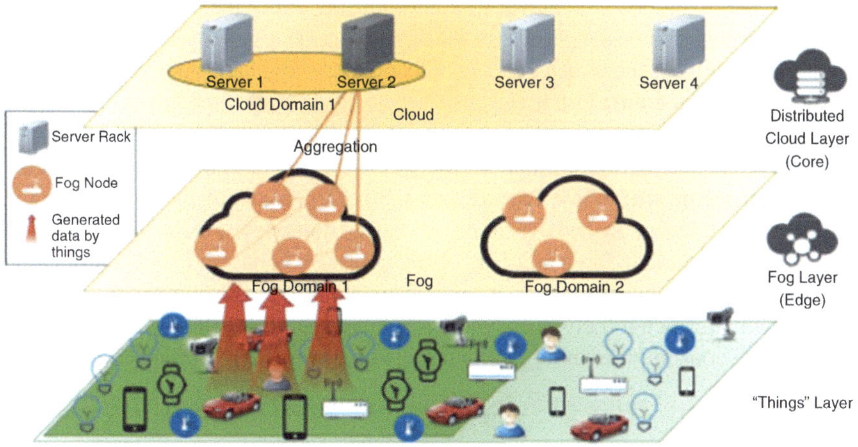

Fig. 8.1 Framework for IoT-fog-cloud architecture

8.2.2 *Fog Computing Challenges*

Scalability: IoT devices are increasing with a rapid rate, which is generating huge amounts of data day by day. It requires more storage capacity and computing capability. Fog server should be capable to grant sufficient resources to support the IoT devices. The main challenge of a fog server will be to compute and respond to all the requests from all IoT devices.

Complexity: Different manufacturers design the IoT devices and actuators. Selecting the best device for sensing is extremely complicated particularly with diverse software and hardware specifications and personal requirements. In the case of highly secured application, it requires particular protocol and hardware which amplifies the inconvenience of operations.

Heterogeneity: IoT devices are designed and manufactured by different vendors. Every IoT devices and sensors have different computing power, storage capacity, sensing capabilities, and radiofrequency. The coordination and management of networks for such diverse IoT devices and the choice of suitable resources are highly challenging.

Latency: Latency is the main concern in cloud computing. The main reason for implementing the fog nodes closer to the IoT devices and in between the cloud a server is low latency, particularly for time-sensitive applications. Latency depends on various factors in processing various services or applications. The fog with high latency will not serve the purpose. Latency is the main concern behind fog computing.

Security: Security is also an important concern in any architecture. Fog nodes must be protected by using various policies. The security aspects of fog computing must not be less than the cloud server as compromising either the physical security or cyber security can be dangerous for the fog system. The security aspects cannot be applied as it is for the cloud, because both cloud and fog have different scenarios.

Resource management: Fog nodes are the intermediate network layer operational with computing power and storage capacity. These devices must match the resource capacity with the cloud servers as the fog layer has low storage capacity.

Energy consumption: While focusing on the other aspects of the fog computing like as low latency, the task has to be offloaded from one fog node to the other. In the fog environment, large numbers of fog nodes are connected, and computation is done in a distributed manner. It can consume more energy compared to a cloud environment when it focuses on resource management and latency. The reduction of energy consumption in fog computing is becoming an important challenge.

8.3 Literature Review

The authors in (1) proposed a novel framework to optimize the configuration of fog network. The objective of this chapter is to minimize the latency problem to allocate the computational resources of the mobile edge servers. Moreover, the authors in (2) presented an algorithm for minimizing maximum delay. The dynamic bandwidth scheme can be designed for minimizing latency.

The researchers have shown the proposal of fog computing, a freshly emerged paradigm that extends cloud computing to the sting of the network. We have a propensity to expand a methodical algorithm to analyze the facility delay tradeoff issue within the fog-cloud automatic data processing system. Researchers formulate the employment allocation drawback, and some fester the primitive drawback into three subproblems, which can be severally solved at intervals corresponding to the subsystems. Simulation and numerical results of fog computing proves that it empower the cloud computing instead of complementing. We tend to hope that this pioneering work will give direction-finding on discovering out the interaction and cooperation between the fog and cloud. The main focus of this chapter is to optimize performance [3].

Researchers have done the comparison between fog computing and cloud computing in detail, illustrating most of the research challenges in fog computing [4]. The following challenges are defined such as security, privacy, node energy, data management, and data consistency traditional desktop and server machines [5].

A fog computing system model is constructed by using a graph theory approach and cloud computing. And on this basis, the researcher proposes a fog computing dynamic load balancing mechanism supported by graph repartitioning. The simulation results verify that the designed algorithm is effective and may be applied to the dynamic load balancing mechanism within the fog computing system. On this basis, the focus of subsequent steps is to enhance. The load balancing algorithm supports the change of the system and therefore the characteristics of the graph repartitioning and further improves the performance of the dynamic load balancing mechanism [6].

This chapter compares the service delay in three diverse modes of the IoT-fog-cloud operation. In no fog processing mode, the sensor nodes process their own requests or send these requests to the cloud. In all fog processing and lightweight fog processing modes, the sensor will send the requests to the fog or cloud; no request will be processed at the sensor end, as the sensor node will work as a sensing device [7].

This study has proposed the model of fog computing in both processing and networking aspects. The model is considered for fault tolerance and privacy parameters for optimization methods. The proposed framework evaluates under a given application scenario and a genetic algorithm combined with a distribution sampling approach [8].

FogTorchΠ is estimated for resource consumption within the Fog layer, which may be wont to minimize the exploitation of certain Fog nodes with reference to counting needs of users. The potential of FogTorchΠ has been illustrated by discussing its application area for agriculture fog application and QoS features [9].

The author has proposed a dynamic resource allocation policy for fog computing-supported Priced Timed Petri nets to enhance the competence of fog resource utilization and improve the quality of services. The algorithm was projected to forecast task completion time and availability of recourses. Simulation results show that the anticipated policy can provide proficient resource selection for user's autonomous task scheduling and improve the use of fog resources [10].

The researchers have proposed a static fog node. The algorithms support optimal storage capacity depending upon the local location requirement and geographical area [11]. In the case of mobile fog nodes, algorithms that move fog nodes as per requirement provide required space and processing capacity [12].

The researchers have given a model for processing the actuator's request in the fog system and systematical model to minimize the service delays within the fog scenarios. The researcher shows how offloading policies can be beneficial for minimizing the delay tradeoff for the actuators. Diverse results are shown to support the claims that changing the parameters can affect the service delay. If we make a change in parameters, the service delay can be minimized in specific scenarios [13].

The resource allocation policies for different task depends on sensors in a shared fog network. In this study, the researchers remove the impractical settings in existing works and create a task offloading model that captures more details of a real-world fog network. Based on the problem-specific analysis, an efficient and exact solution method is proposed as a replacement offloading Scheme [14].

In this chapter, the authors have focused on minimizing the latency of the model while inserting the tasks on fog nodes. The author has proposed an exact solution for the placement of the task on the desired fog node. The study was achieved using the iFogSim simulator fog environments respectively. The proposed solutions endeavor to discover the preeminent decision placement for each module [15].

The authors have proposed a VANET system for smart cities. In this system, the information is gathered by the roadside assistance units (RSU), and after processing and verifying, the information is transmitted to the local network units that are fog system. Fog system further that desired information to the vehicles which are the part of the system. Specified clusters are created of a specific geographical area of the city [16].

The aim of this study is to reduce the load on the cloud server. Fog layer is introduced in the IoT-cloud architecture not as a replacement for the cloud nodes; it is an extension in computing and storing capability of the cloud architecture. Fog computing has many open challenges, and issues still exist, such as resource management issues. The proposed framework has the ability to achieve permanent network patterns and highlights the important benefits of fog in the computing ecosystem [17].

This chapter concludes that the task offloading techniques must be hidden from the application developers. The data shared with the fog node may be privacy-preserving so the data must be kept confidential and must be shared by the users who are the part of the system. A privacy-preserving algorithm must be applied to maintain privacy. It also keeps track that in maintaining privacy the performance must not be compromised. A comprehensive SLR was provided on the resource management approaches within the fog computing environment [18].

This chapter discussed a fog computing-based traffic light control strategy and traffic light control architecture. The smart traffic light system gathered data from the vehicles. The vehicles work as the sensing nodes. The data is collected in the form of the velocity of vehicles, density, speed, and timestamp. In this architecture, the phase timing tasks for one intersection are often handled by an area fog node in real time, and regional optimization tasks are going to be left for the centralized cloud. SUMO system has been used to track the long queue on the traffic signal and transfer these signals to the local fog system. Further after processing the task, the local fog system transmits those signals to the local traffic lighting system [19].

The research addresses the problems of task data offloading in fog computing considering different delay deadlines for the tasks initiated by the end users. The main objective of task offloading is to obtain the load balancing between the neighbor fog nodes. The best neighbor node will receive the task for processing. The selection of the node will depend upon certain parameters as the physical distance between the nodes must be less and the selected node must not be busy in other tasks. There must not be a long waiting queue of the task. The transmission delay is the data transmission rate in a real-time environment [20].

The chapter studied the task of offloading for real-time fog systems by optimization of partial task offloading decisions, storage, and resource allocation [21]. Offloading the task from one fog node to another fog node can be a time-consuming process, and if most of the local fog nodes are busy, it will take more time to check and offload. In such cases when any of the fog nodes is not free, the task is transferred to the cloud system [32]. The main objective is to reduce latency. If the fog node does not meet the minimal energy consumption requirement, the fog network model will not be considered as a successful model. A joint optimization task offloading technique can be considered for such scenarios [22].

The number of IoT devices is increasing rapidly, and they are producing a huge amount of data. So, cloud computing is unable to meet the real-time requirement. This chapter has designed an optimal job scheduling algorithm to minimize the latency for time-sensitive applications. The proposed algorithm schedules jobs on fog devices on the basis of length and reduces the average loop delay and network usage. The proposed SJF algorithm reduces the common waiting time [23].

8.4 Smart City Ware Architecture

Transmitting the entire city traffic data to the centralized cloud server will overburden the cloud. There is no need to send the entire city local traffic data to the centralized cloud. This local data can be processed through the localized server. The main objective of SmartCityWare architecture is to develop and deploy a virtual environment for a smart city. SmartCityWare architecture can be implemented for a variety of services. These services are classified into core services and environmental services. Core services are designed specifically for the core operations of SmartCityWare, such as the broker, security, service invocation, and location-aware services [24, 33].

One of the application areas can be an intelligent traffic light control system for the smart city. This intelligent traffic light control system will manage, control, and monitor the city traffic [25, 26]. It will also forecast the traffic pattern according to the previous routine. SmartCityWare will monitor the daily traffic, manage it, and if required will send the alerts to the control room accordingly [27]. The processing and information transferring will be localized. This application can be an add-on vehicle-to-vehicle and vehicle-to-infrastructure communication. This system will minimize the vehicle's travel times, decrease the traffic delay, and improve vehicle average velocity [28, 29]. With the help of such a system, emergency movement vehicles can be prioritized [30, 31]. In case of a vehicle collision or accident, information can be transmitted to other networks to the city so that the traffic can be diverted to other areas. An alert can be sent to the city medical team so that medical help can be sent to that affected area [24].

8.5 Research Gap

- There is a need to reflect on the supplementary aspect of IoT requests, like the quantity of data that every request can carry. That might be helpful to examine delay, cost, and energy tradeoffs in fog offloading schemes.
- Fog computing requires energy consumption and latency reduction for the end users' device when the task offloading is facilitated by a cluster of fog nodes.
- It is more required to consider possible ways in fog computing for network bandwidth constraints.
- There is a need to study and minimize the delay-cost tradeoff in fog computing.
- Study the effect of different objective functions on the efficiency-fairness tradeoff; efficiency means overall performance, and fairness is related to the individual performance achieved by the task.
- For improving the overall network performance, there is a need for the optimization of multiple objects.
- Placing data in fog computing can produce high overhead in terms of storage cost, network traffic energy consumption, and network traffic in case of more replica of fog.

8.6 Propose and Execution

Several requests of fog computing are latency-responsive, at the same time as additional ones are delay forbearing. The workloads caused by these submissions are dynamic, of variable length, and from time to time require priority execution in cooperation with edge and cloud. Claims contend for limited wealth of assorted devices in heterogeneous situation. These workloads are handed over and performed at a choice of fog nodes. If round-robin and first come first serve job scheduling algorithms for job scheduling in fog computing are used, it provides identical right of way to each and every one job, which results in increased rejoinder time for jobs with little burst times. Nevertheless, the aim of fog computing pattern is to diminish waiting time, response time, and network traffic.

For that reason, at hand is to require proposing and executing a job scheduling algorithm in fog by way of pursuing the following objectives:

- Minimize latency (request loop delay).
- Resourcefully make use of the resources of fog devices.
- Reduce the network tradition.

8.7 Performance Metrics

Researchers proposed three metrics that are loop delay, power utilization, and network usages.

- **Average loop delay:** A control loop to calculate the end-to-end latency of all modules in the loop. To compute the loop delay, we compute the average CPU time and CPU utilized by all nodes.
- **Power utilization:** The power of any fog device can be computed by the power of all the hosts in a set period of time for execution.
- **Network usages:** Increasing the number of devices increases the network usage, and it results in network congestion. This congestion eventually effects in poor performance of the request running on cloud network. Fog computing helps in diminishing the network congestion by distributing the load on intermediate fog devices.

8.8 Conclusion

The rapid increase in IoT devices is bringing about huge amount of data. Due to this rapid growth, cloud computing is not able to meet real-time requirements such as mobility support, location awareness, and low latency. To overcome such restrictions, fog computing has materialized a new computing model that consummates cloud computing by providing support for real-time analytics, processing, and storage services closer to the IoT devices. The IoT devices have limitations of computing and storage. Fog nodes have the computational capability, storage capacity, and networking resources for IoT devices. Here is a necessity to reflect on the accompanying characteristic of IoT requests, like the amount of data that every request can carry. That might be helpful to inspect delay, cost, and energy tradeoffs in fog offloading schemes. Fog computing is calling for energy consumption and latency decline for the end users' device when the task offloading is facilitated by a cluster of fog nodes. Put the local request on the fog rather than the cloud which reduce the power consumption and delay. Localized information is required, processed, and transmitted through these localized fog nodes on a real-time basis. This chapter concludes that the placement of IoT modules and job scheduling on fog nodes will reduce the overall latency and energy consumption.

References

1. Yousefpour, A., Ishigaki, G., Jue, J. P. (2017). *Fog computing: Towards minimizing delay in the Internet of Things*. Proceedings – 2017 IEEE 1st International Conference on Edge Computing EDGE, pp. 17–24.
2. Wang, S., Zhao, T., & Pang, S. (2020). Task scheduling algorithm based on improved firework algorithm in Fog Computing in IEEE Access, 8, 32385–32394. https://doi.org/10.1109/ACCESS.2020.2973758.
3. Atlam, H., Walters, R., & Wills, G. (2018). Fog computing and the internet of things: A review. *Big Data and Cognitive Computing, 2*(2), 10.
4. Peng, M., Yan, S., Zhang, K., & Wang, C. (2016). Fog-computing-based radio access networks: Issues and challenges. *IEEE Network, 30*(4), 46–53.
5. Souza, V. B. C., Ramirez, W., Masip-Bruin, X., Marin-Tordera, E., Ren, G., & Tashakor, G. (2016). *Handling service allocation in combined Fog-cloud scenarios*. 2016 IEEE International Conference on Communications ICC 2016, pp. 0–4.
6. Deng, R., Lu, R., Lai, C., Luan, T. H., & Liang, H. (2016). Optimal workload allocation in fog-cloud computing toward balanced delay and power consumption. *IEEE Internet of Things Journal, 3*(6), 1171–1181.
7. Ningning, S., Chao, G., Xingshuo, A. N., Qiang, Z., Ghylfh, H., Dv, V., et al. Fog computing load balancing, pp. 156–164.
8. Xu, X., Fu, S., Cai, Q., Tian, W., Liu, W., Dou, W., et al. (2018). Dynamic resource allocation for load balancing in fog environment. *Wireless Communications and Mobile Computing, 2018.*
9. Brogi, A., Forti, S., & Ibrahim, A. (2017). *How to best deploy your fog applications, probably*. Proceedings – 2017 IEEE 1st International Conference on Fog Edge Computing ICFEC 2017, pp. 105–14.
10. Ni, L., Zhang, J., Jiang, C., Yan, C., & Yu, K. (2017). Resource allocation strategy in fog computing based on priced timed petri nets. *IEEE Internet of Things Journal, 4*(5), 1216–1228.
11. Taneja, M., & Davy, A. (2017). *Resource aware placement of IoT application modules in Fog-Cloud Computing Paradigm*. Proceedings IM 2017–2017 IFIP/IEEE International Symposium on Integrated Network and Service Management, pp. 1222–1228.
12. Moysiadis, V., Sarigiannidis, P., & Moscholios, I. (2018). Towards distributed data management in fog computing. *Wireless Communications and Mobile Computing, 2018*(i).
13. Yousefpour, A., Ishigaki, G., Gour, R., & Jue, J. P. (2018). On reducing IoT service delay via fog offloading. *IEEE Internet of Things Journal, 5*(2), 998–1010.
14. Mukherjee, M., Shu, L., & Wang, D. (2018). Survey of fog computing: Fundamental, network applications, and research challenges. *IEEE Communication Surveys and Tutorials, 20*(3), 1826–1857.
15. Nikolopoulos, F., & Likothanassis, S. D. (2018). On the move to meaningful internet systems. OTM 2018 conferences: Confederated international conferences: CoopIS, C&TC, and ODBASE 2018, Valletta, Malta, October 22–26, 2018, PROCEEDINGS, Part II [Internet]. Vol. 11230, On the move to meaningful internet systems. OTM 2018 Conferences. Springer International Publishing; 2018. 1–11 p. Available from: https://doi.org/10.1007/978-3-030-02671-4_6
16. Vidyasankar, K. (2018). *Distributing computations in fog architectures*. In Proceedings of annual ACM symposium on Principle of Distributor Computing, pp. 3–8.
17. Al-khafajiy M, Baker T, Asim M, Guo Z, Ranjan R, Longo A, et al. COMITMENT: A fog computing trust management approach. Journal of Parallel and Distributed Computing [Internet]. 2020;137:1–16. Available from: https://doi.org/10.1016/j.jpdc.2019.10.006.
18. Ghobaei-Arani, M., Souri, A., & Rahmanian, A. A. (2019). Resource management approaches in fog computing: A comprehensive review. *Journal of Grid Computing, 18*, 1–42.
19. Tang, C., Xia, S., Zhu, C., & Wei, X. (2019). Phase timing optimization for smart traffic control based on fog computing. *IEEE Access, 7*, 84217–84228.

20. Mukherjee, M., Kumar, S., Shojafar, M., Zhang, Q., & Mavromoustakis, C. X. (2019, May). *Joint task offloading and resource allocation for delay-sensitive fog networks*. IEEE International conference communications, pp. 1–7.
21. Li, X., Liu, Y., Ji, H., Zhang, H., & Leung, V. C. M. (2019). Optimizing resources allocation for fog computing-based internet of things networks. *IEEE Access, 7*, 64907–64922.
22. Mukherjee, M., Kumar, S., Zhang, Q., Matam, R., Mavromoustakis, C. X., Lv, Y., et al. (2019). Task data offloading and resource allocation in fog computing with multi-task delay guarantee. *IEEE Access, 7*, 152911–152918.
23. Jamil, B., Shojafar, M., Ahmed, I., Ullah, A., Munir, K., & Ijaz, H. (2020). A job scheduling algorithm for delay and performance optimization in fog computing. *Concurrency and Computation, 32*(7), 1–13.
24. Mohamed, N., Al-Jaroodi, J., Lazarova-Molnar, S., Jawhar, I., & Mahmoud, S. (2018). *A service-oriented middleware for cloud of things and fog computing supporting smart city applications*. 2017 IEEE SmartWorld, ubiquitous intelligence & computing, advanced & trusted computing, scalable computing & communications, cloud & big data computing, internet of people and smart city innovation SmartWorld/SCALCOM/UIC/ATC/CBDCom/IOP/SCI 2017, pp. 1–7.
25. Bittencourt, L. F., Diaz-Montes, J., Buyya, R., Rana, O. F., & Parashar, M. (2017). Mobility-aware application scheduling in fog computing. *IEEE Cloud Computing, 4*(2), 26–35. https://doi.org/10.1109/MCC.2017.27.
26. Sarkar, S., & Misra, S. (2016). Theoretical modelling of fog computing: A green computing paradigm to support IoT applications. *IET Networks, 5*(2), 23–29. https://doi.org/10.1049/iet-net.2015.0034.
27. Du, J., Zhao, L., Feng, J., & Chu, X. (2018). Computation offloading and resource allocation in mixed fog/cloud computing systems with min-max fairness guarantee. *IEEE Transactions on Communications, 66*(4), 1594–1608. https://doi.org/10.1109/TCOMM.2017.2787700.
28. Li, S., Zhai, D., Du, P., & Han, T. (2019, February 1). Energy-efficient task offloading, load balancing, and resource allocation in mobile edge computing enabled IoT networks. *Science China Information Sciences*. Science in China Press. https://doi.org/10.1007/s11432-017-9440-x.
29. Shi, W., Cao, J., Zhang, Q., Li, Y., & Xu, L. (2016). Edge computing: Vision and challenges. *IEEE Internet of Things Journal, 3*(5), 637–646. https://doi.org/10.1109/JIOT.2016.2579198.
30. Chen, S., Zheng, Y., Lu, W., Varadarajan, V., & Wang, K. (2020). Energy-optimal dynamic computation offloading for industrial IoT in fog computing. *IEEE Transactions on Green Communications and Networking, 4*(2), 566–576. https://doi.org/10.1109/TGCN.2019.2960767.
31. Jiang, J., Tang, L., Gu, K., Jia, W., & Sgandurra, D. (2020). Secure computing resource allocation framework for open fog computing. *The Computer Journal, 63*(1), 567–592. https://doi.org/10.1093/comjnl/bxz108.
32. Gupta, S., Vyas, S., & Sharma, K. P. (2020, March). *A survey on security for IoT via machine learning*. In 2020 international conference on Computer Science, Engineering and Applications (ICCSEA) (pp. 1–5). IEEE.
33. Kumar, A., Kumar, P. S., & Agarwal, R. (2019). *A face recognition method in the IoT for security appliances in smart homes, offices and cities*. 2019 3rd international conference on Computing Methodologies and Communication (ICCMC). IEEE.

Chapter 9
Changing World: Smart Homes Review and Future

Pooja Tiwari, Vikas Garg, and Rashmi Agrawal

9.1 Introduction

Smart homes are a very unique and a futuristic profitable idea which gives the user a feeling of living in a comfortable environment, a luxurious way and a high standard of living. It also includes various modern technologies which give the smart home the best environment to live in and provide all the luxury to the user. It is believed that soon this industry will grow and eventually the demands and the needs of these homes will boom. Giving comfort and an unpredictable experience to the user is the special feature of these homes, which are said to be smart homes and the homes of the future world. Easily accessible to all the smart features of the homes and by just one click and one control which is in your hands itself? By 2060, the majority of the population across the world will live to around 60 years on an average [1]. There is a possibility that people in this age bracket will have to handle the different problems in independent living and quite possible that they will be more prone to long-term chronic diseases. According to the World Health Organization (WHO), it has been concluded that across the world almost 650 million people are living with a disability. The major reason which are identified for the disability includes chronic diseases such as diabetes, cardiovascular diseases and cancer and injuries due to road traffic accidents, conflicts, falls, landmines, mental impairments, birth defects, malnutrition and HIV/AIDS and other communicable diseases. Certain alternative has to be pondered upon as it is quite difficult to provide

P. Tiwari (✉)
ABES Engineering College, Ghaziabad, India

V. Garg
Amity University Uttar Pradesh, Noida, India
e-mail: vgarg@gn.amity.edu

R. Agrawal
Manav Rachna International Institute of Research and Studies, Faridabad, India

M. Moh et al. (eds.), *Smart IoT for Research and Industry*, EAI/Springer Innovations in Communication and Computing,
https://doi.org/10.1007/978-3-030-71485-7_9

continuous medical supports to these patients for an uncertain period of time. The solution of this problem can be provided if we can accommodate both healthcare services and assistive technologies in the homely condition of the individual [2]. It is important to choose the location wisely not just because to create smart homes but to give the actual comfort and that luxury which the smart home user wants. Everything needs to be perfect from one end to another, connecting all the dots. A user-friendly and robotics-friendly home is a best way to increase the human and robot conversations as well. Comfort is the important aspect of these types of homes, and that is why the user wants to buy it as well. Smart homes will not be a luxury in the coming times; moreover it will become the necessity to the entire user, a big reason to buy smart homes then. A comparison between various smart home protocols, let's say modern examples such as Sigsbee and Wi-Fi, is also provided. We cannot say that smart homes are really a need of the hour and users need it because of various security issues linked with smart homes and its applications. Many kinds of research show that up to certain extent smart homes security can be really good but once it starts showing security issues and start making a threat to the user. Many services need to be assembled and arranged in that order so that smart homes could be not just user-friendly in terms of comfort and luxury but also offer security as well. Smart homes may possess a good technology, but still many loopholes are there, and it needs to be rearranged for the better enhancement of smart homes and better comfort which the smart home user wants from these types of homes. The world needs smart homes in a protected and a secured way [3]. It can be said that smart homes are a need of the modern world towards a more comfortable and luxurious life, but the application installed should be 100% safe and non-hackable. Some major drawback needs to be considered, so that this idea of smart home can be moulded into the flexible projects like the normal. The digital world will totally be a different world, and everything is becoming digitised, controlled by machines and automated systems. Smart homes are one of those concepts which are taking you to the feeling of high-tech homes. Smart home is universal computing in which ambient intelligence technology is used and which helps in monitoring the home environment. Smart homes and the different technologies associated with them need to be explored. There are many real estate companies which are promising that smart homes will provide significant amount of domestic comfort, security, expediency and relaxation while on the other hand also minimising the energy used due to optimisation of home energy management. Smart homes have potential to achieve multiple goals, but it mainly depends upon how the various facilities are used by the householders. As we are growing and becoming modernised, our needs and demands are also increasing towards the best homes which offer us more comfort, security, convenience and luxury. It has given the description about the combined evidence about sensors, different systems and communication protocols and multimedia devices which are extensively employed in the implementation of smart homes. Smart homes not only give an idea about the future of homes, but on the nature part as well, it is said to be very beneficial. These homes are designed in those ways which are sustainable and fit to the needs of future and present scenarios.

9.2 Definition of Smart Homes

A smart home is an application based on various technologies which AI as well and it is of the newest-to-newest technologies are implemented in smart homes in terms of better comfort levels, best lighting. Smart homes and their first explanation or definition was firstly proposed by Lutolf. According to Lutolf, "the smart home concept is the integration of different services within a home by using a common communication system. It assures an economic, secure and comfortable operation of the home and includes a high degree of intelligent functionality and flexibility." There were several things by which its definition got influenced. For example, according to Berlo, a smart home is something which is totally dependable upon automation and where everything is controlled automatically. There were many more definitions proposed by Winkler and others. But the one that is common to all the definitions is that everyone states that a smart home is technology driven which is based on automation. There are various researches done by the UK Department of Trade, and according to those, a smart home is a combination of all the networks and communications allowed by several applications which are controlled, automated and monitored [4].

A smart home should consist of the following:

Network connectivity which includes wireless connectivity, cables and wires
Gateway system which can manage featured system and should possess intelligent system
Automated home: Products and services linked to the system within and outside the premises of the home

UK research gives more accuracy about smart home and its intelligence and how it is totally dependable through remote access. Satpathy's study also gives the best definitions to smart and intelligent homes. According to Satpathy, "a home which is smart enough to assist the inhabitants to live independently and comfortably with the help of technology is termed as smart home. In a smart home all the mechanical and digital devices are interconnected to form a network, which can communicate with each other and with the user to create an interactive space". The definition was not mentioned and stated by the author. There are many trends in the automated homes, and there will be upcoming technology which can amaze the future concepts of living [5].

9.3 Smart Homes and Their Users

Technology-driven homes [52] and the various researches have given the actual challenges for smart homes and its surrounding. According to current factors and various studies, these smart homes are not still user-friendly due to many reasons, like technology reasons which today even many are not aware of. As of various reviews and studies, it is found that the demand for smart homes by the users is continuously growing, and due to its high demand, the industry is also growing much faster. Smart homes and its associated challenges were figure out and came from many researches are consoled below:

- The technology should be able to fit with user needs.
- SHT and its ease should be administered and needs are to be considered.
- Interoperability between systems.
- Consistency.
- Confidentiality and safekeeping homes or support to religious or pro-environmental values.

Summarising this we can say that the concept of smart homes represents that it should be technological based with the essences of a calm and peaceful life.

The smart home should be flexible enough to adjust and maintain the flexibility with the user and should fit with the day-to-day changing activities of the user. The automation and the technology should be smart enough to interact with the user. Technology should be flexible enough to negotiate with the user with the needs and the wants [6]. SHTs should possess the feature of adapting them with the challenging and complex mixed messages which they are receiving from the multiple sources. Third, SHTs must not overpower their users by using too much control to manage their smart home. The user doesn't want to overpower the options of smart homes because all they want is their daily tasks and activities to be performed easily and peacefully: interaction of "human–home collaboration". This requires systems that allow users to communicate with these SHTs so that the user also gets a sense of interaction with the technology and does not make it much complicated in terms of its technology and its related knowledge and learning.

9.4 Domestication of Technologies

Smart homes and its domestication technology are quite essential to make it more profitable and reliable to all. Early research on the diffusion of several information and innovations which attribute at its cost and the benefit it is giving to its user is endless. Seamless communication factors and barrier-free technology transmission are the major benefits of the smart home. Various domestications of technologies are seen over a time frame, but still gradual changes and moreover speedy changes are required, and they need to be also diffused through various communities and will eventually put a negative role for users "who simply adapt to what is offered to them" [7]. The basic concept involves emphasise various technologies in 'taming' 'wild', within various technologies are combined to make it functional with smart homes. Specifically, domestication requires that users undertake three types of work:

- *Cognitive*: the work related to attaining the knowledge regarding technology and its implementations.

- *Practical*: the work related to knowledge attainment about the usage of technology.
- *Symbolic*: the work related to learning about and constructing the meaning of the technology and how to incorporate it in identities.

Domestication stresses how, through this work, technologies and their users co-evolve as technologies enable new routines and identities and are thus given particular functions and meanings. Locally domestication is rarely a "harmonious process". The various issues and their related conflicts cannot be ignored, up to some these technologies are very and useful and it's implemented but after that, some are there which have no specific use as such. The domestication process is never complete. The technologies in the field of smart homes are always developing, and speedy changes are there. Thus, we can say that we can never assume a smart home with some fixed technology; it needs to be changed and upgraded from time to time. Innovation processes and new driven technologies are really a need of the hour, and it needs to be focused on if we need to grow in the sector of smart homes, its technologies and its various applications [8]. Technologies with different functions have different sets of functions, and some of them need to be innovated for future technologies as well. Domestication and its various processes are applied in the varied sectors of energy development, for example, renewable energy and its small scale of production as well. It depends on how the user uses the SHT, because sometimes specific knowledge and learning is required to get these SHTs to be understood and then put into action. These smart homes are basically a combined mixture of present as well a touch to future technologies as well.

9.5 Smart Home Projects

In the past two decades, the concept of smart homes have evolved, and it has also communicated many ideas, purpose, features and values. Smart homes and its usage are of varied domains, and no doubt it is specialised in many domains of living by giving comfort and luxury. These types of homes are basically designed to give the best environment the user could have and ever imagined [9]. Many researches have proved that these types of homes are basically network and technology driven which are specialised in each area of its segment. Like if we talk about communication, then it has special sensors to navigate the best communication and many more advanced features one could think the smart home will give, and it is actually customisable according to the needs and demands of the user.

One of the key reasons why the user opts for smart homes is because of their luxurious and comfortable style. The comfort which the user gets from such types of homes is unimaginable to many and the user can adjust the environment of the home according to the needs at a specific moment, lighting at special times, best view, connectivity, and automated driven technology equipped home to provide world-class services with each and everything automated and customised according to the user needs. Sensors are there to track and monitor various parameters like health, heart rate, sleep pattern and many more.

9.5.1 Activity Identification and Event Automation

These smart homes are totally different from the ordinary homes, and they have a very different concept from ordinary homes. These homes are specially designed to give the best comfort to each its users in each way and in each segment [10]. The home-based on full automation and the best technology is driven with each of it creating with artificial intelligence and some under robotics as well. Many leading universities have done various researches on smart homes and its future to the users. The architecture of this house is specially designed in many layers, and four of them are communication, physical, information and decision; around all these four factors the working of the smart home revolves. These types of special arrangements require lots of connectivity and high engineering skills, because the technology is upgrading at each day and at each moment. These types of special homes are equipped with not only upmost technology, but it also comes with flexible environment and flexible changes that the users want to do according to the needs. These are totally customisable homes with high-tech security from in to out. Many models have taken into account smart homes like the Core Lab, the name given to smart homes only with all the specialised technology-driven and friendly use. These types of smart homes models are not just made to advance the technologies but to start the more interaction between the human and the automation technologies [11].

This research is an excellent example of a smart home with distributed intelligence and its importance in the future era.

9.5.2 Remote Access and Control

Smart homes and its accessibility in your hands. Sounds interesting? But it's actually true; users can turn into anything via the Internet or Wi-Fi. These kinds of projects permit the user to have remote admittance and supervise and control the environment of the home. In the present scenario, the Internet has become one of the most common and pervasive technology which enable the two-way communication between the home and the user. Many a times for this purpose of communication, telecommunication infrastructure is used. Some projects are mentioned further through which it is possible for the user to control and supervise the home appliances from remote locations by direct interactions. In such circumstances, only the platform for the remote access and control is created, and it does not possess intelligence [12]. Perumal et al. from the Institute of Advanced Technology of Universiti Putra Malaysia (UPM) have introduced a design for smart homes which implements Simple Object Access Protocol (SOAP). In smart homes for the purpose of interoperation of home appliances, a control module is created which is referred to as SOAP. In the residential management system, almost 15 feedback control channels were created which are web based in nature. In case the server does not perform well, this system can also be controlled remotely by the help of module of SMS. It provides the benefit of complete, real-time control which is a bi-directional and monitoring system. Relay-based switch system has been implemented, and it doesn't follow any standard for the protocols in communication. Wang et al. have put forth the concept of smart home having monitoring and controlling systems. By the help of an embedded controller, these kinds of systems can be controlled from remote locations. Various types of GUIs and PCs are developed by the authors. There is a unique address for each device. Each device has a unique address. To control the device, a new command format has been introduced. Although in the present scenario the protocol which is existing is sufficient, still a new protocol with a new command format has been proposed [13]. The architecture is a peer-to-peer model based on multiple OSGi platforms. Their future works include security at many levels which can be easily attainable in these types of homes with some more modification in technology.

9.5.3 Healthcare

Smart homes provide healthcare facilities for patients, elderly people, and healthy people. Healthcare services are one of the major concerns for everyone, and it is more than a concern; it's everyone priority, and if smart homes are well equipped with such technologies, then it will be great relief to many [14]. These types of smart homes are really beneficial in providing instant health and checkup report, which in turn is really helpful for the elderly people. Remote healthcare system is one of the essential and important features of smart homes, and it can get the instant service for emergency support.

9.5.4 Local Monitoring

Smart homes could be really beneficial in terms of monitoring one's health and physique, and it accurately monitors tall over health of an individual, and it covers the untouched portion of the examination as well. Based on the various technologies and their intervention in human health could be very beneficial, and in these where health and maintaining physique is everyone's priority. It accurately monitors your heartbeat count, the cardiac cycle, the 24/7 health app you can say it is like that only, Covering all the important aspects of human health. It will eventually be important for monitoring the health status of the elderly people. It helps to detect various disorders which might occur in the human body and which the user may not be aware of like sleep disorders, sleep disturbance and lack of sleep, blood pressure, heart risks, and change in sleep patterns. It eventually notices the health status of the individual and generates the report based upon that which determines the status of individual health [15].

From health to overall development, these smart homes are playing an effective and an efficient role in the lives of an individual. These smart homes could be the homes of the future, and we can also say this as the 24/7 doctor of each user of these smart homes. From luxury to necessity, these smart homes will adapt like this in the coming future, and no doubt it will not be a bad deal. In the way of exploring, these researchers have created a simulation room for SELF, which includes the sensor array for blood pressure on the bed, washstand having a display, and a ceiling lighting dome with a microphone. So, this system can be considered as system of self-assessment which is utilised for the evaluation of human health. The end time is estimated using a third-order transfer function by modelling the foot pressure on the floor. By adopting the process of video-recorded data, authors have tried to determine the error in the case of end and start time. In the case of Si-St transfer, the other features of the system cannot be determined, for example, in the case of the use of hands to assist in the transfer, a measure of the leaning of the forward trunk, frequency of unsuccessful transfers, and a measure of stability while standing. Routines can easily be recognised by the help of developed systems and also deviations form routines. It can provide the user the basic information with respect to their daily rhythm, sleeping disorders, and medicine taking time. It works sensibly for lighting control and its optimisation according to the environment as well [16].

9.5.5 *Remote Monitoring*

Remote monitoring is the special application of smart homes which enables the user to get any medical assistance or medical support in any emergency situation. The home monitors the overall physiological aspects of the human body through several sensors installed in it. The sensors are basically specialised application which are automated and work on its own for the specific function. Each sensor is designed for each and specific function it was designed for, for example, heat sensors to detect the heat powers, thermal sensors, IR sensors and many more which are installed in smart homes under the application of remote monitoring.

The network policy of these smart homes is very efficient and not disrupting the user to change its older networks. It will continue upon the same with some certain and required additional changes [17]. Providing lower-cost solutions and collection of the data, from the user existing or the previous telecommunication network, not depending upon which type of network it is and what is the location. Weather it is LAN or WAN, it doesn't matter.

Their core component includes a set of various sensors in the involvement of these smart homes, a sensor bus, an intelligent monitoring system and a control unit. For the purpose of recording different types of data such as real-time, event, command and control, an integration communication network is created. This network is an integration of home LAN and body area network (BAN). It has a distributed intelligent system in the form of smart sensors, smart therapy units, body hub, local intelligence unit (LIU) and client's healthcare record (CHR). To help the elderly and disabled individual, telemedicine system is used. It consists of physiological sensors (pulse oximeter sensor and the blood pressure sensor) and general-purpose sensors (weight sensors, motion detectors and light sensors). The basic or we can say the overall interaction still depends on human interaction for medical report analysis and its checking.

9.5.6 *Security*

There is possibility of security threats in the case of smart homes. Majority of the problems which are related to security threats are weak users and authentication schemes of different devices. These attacks on security can be both remotely and generally located. Certain projects which have a possibility of security threats are further discussed in the study. It can be concluded by the help of different researches that most of the security mechanisms are taken from the computer security that are existing [18, 19]. The prime objective of the gateway is the implementation of user authentication scheme. Based on the authorised user control, it is possible for the individual user to login and further access the billing information. This entire system is fully equipped with firewall and virus protection software. Based on the functionality of the product, another crucial part of this study is the illustration of different taxonomy which are related to the taxonomy of the common security. These notions are majorly dependent upon the previous experience related to security of the computer network which includes user impersonation, device impersonation, service interruption, data alteration, worm/virus, phishing, data wiretapping, firmware alteration and OS/software vulnerability. For the prevention of these security threats, the threats are divided into different levels based on their difficulty along with the proposed solution to handle these problems. During the phase of easy setup, the key of public cryptography is used for the process of device authentication. The security features associated with these special types of smart homes can be dissolved once the technology is friendly to people and future needs.

9.5.7 *Discussion*

Smart homes and its concept totally depend upon the user, how he wants his smart home to look like and what type of customisation he wants in that smart home. In general, smart homes are providing the facility with comfort, safety, security, remote control and energy conservation. Furthermore, it also extends the support for elderly and disabled people in terms of providing the support of healthcare [20]. These are those cases which although suffering from chronic long-term disease but there is not much requirement of critical medical support. It is not possible rather to provide the medical care in a conservative way in some medical centre for a time period which is uncertain in nature. There is also a possibility that some elderly individuals lose their senses and have issues related to memory, and due to which they cannot even afford to go to a hospital. In these circumstances there is a need of safety and security and an immediate requirement of health support. Smart homes can support the disabled, for example, patients with bone fractures, hearing problems, blindness and mental disorders [21].

These set of individuals requires to be monitored in a continuous way in the intelligent environment. Smart homes transfer medical facilities to citizens. There are many patients who can be benefitted from these technologies and by using these smart technologies. Moreover, a smart home is just not a home, but it's a place of comfort and luxury a user wants, benefitting from health and physique to overall body development. A smart home is a network of several communication and technological devices that are connected together to give comfort and luxury which the user wants from these types of homes. More or less smart homes take care of the satisfaction a user requires from its smart homes, all the technology used in these smart homes are user-friendly and both can adapt to each other convenience and as per the usage of it [22]. A smart home enables the user to work in an safe and an comfortable environment with all the benefits required for that. There are still several challenges in the adoption of these smart homes, and we cannot ignore this; this needs to be kept into account so that after countering these challenges, customers can feel at ease in buying smart homes in the future.

Decision support facilities are added in the methods of data processing and algorithms so as to introduce and expand services. The below-mentioned sections will discuss various components which are responsible to manage functionality and efficiency of smart homes. Efficiency of smart homes totally depends on its technological advancements and as well as how the user is educated to use these types of well-equipped technology [23].

9.6 Devices and Equipment

Smart homes are generally a type of home which is fully equipped with latest technologies and advanced features installed in the homes in the form of several automated applications to give the best comfort and luxurious living style to the user.

To identify the different devices which are active some power sensors are employed. To communicate the date consumed by the different sensors are quite low bandwidth. For the purpose of monitoring the patient's certain equipment's are required to measure blood pressure, body temperature, body weight, and heart rate. There are certain equipment which are sophisticated in nature and which are used such as ECG, PPG, etc. [24]. Based on the devices that are utilised, the signals of the bio-format will vary. Multimedia devices have been introduced to create an interactive and robust home environment. Cameras and microphones are common data acquisition tools. Plasma displays, headsets and LCDs create a platform for information exchange. The technology enhances the living environment. Recently, motes have being used in smart homes. A mote is a wireless node used in a WSN (wireless sensor network). It provides an integrated, stand-alone, predefined functionality for the rapid development of a sensing network. Motes are expensive and limited to specific uses [25]. The smart homes are basically managed under various sensors which control each and everything that is happening inside the home as well the movement happening outside the home, creating the best environment for the user. An example is light sensors which are basically useful for illuminating the light at very specific locations giving the accurate or the best results.

There are many more which make the user-friendly with these types of smart homes which include, image processing, audio-video captures and visuals, multimedia effects and sounds, noise filters to enhance the quality of sound inside the home and not letting the outside disturbance come inside. Voice recognition, face recognitions and video processing are some of the advanced features of these smart homes, and within the development of these types of homes, many more technology will come up which will amaze the user to use the smart homes [26, 27].

Network detection sensors and networking sensors are installed in the home for the better and best connectivity with anyone and anywhere, making and giving you the best and accurate catch of network possible in its limits. Sometimes this might too disrupt by not changing and giving the right network locations for better communications. Moreover, there needs to be more quality and enhancement with video processing and media as well.

9.7 Communication Media and Protocols

A smart home is fully equipped with the latest and the best technology one could use to make the home with best comfort and a luxury style. These smart homes have a strong network of connections linked to each other to communicate better and provide better assistance to the user as a friendly technology [28]. The communication media and its protocols are further classified under various segments like wire, wireless and hybrid.

In these types of protocols and special arrangements, some only use wire for communications, while others use both wire and wireless systems of communications. Provide a wide range of data connectivity with strong bandwidth for multimedia data transmissions. The popular Wi-Fi is a must for all the entry to the smart homes, or we can say that all the devices in the home are connected to the Internet or Wi-Fi for seamless functioning and presentation [29]. Moreover, it is for sure that the quality of smart homes is dependable on its communication types and through various channels, it is transmitted, various systems, protocols are there, vendor to user or user to vendor transmission also sometimes seen. But the problem arises when such a type of applications is supported to only specific or runs by specific application and software usage. These are some protocol-based devices [30, 31]. Generally, these types of problems arose when using different merchant products without altering the protocols of the existing product because these are some applications which are supported or casted on some specific set of controls and they'll not perform once the protocol changes and its network coverage changes [32]. But this type of problem can be resolved by some additional expense in changing the protocols as the transitions are quite expensive.

The varied types of interconnections are there which are helping to give the appropriate comfort level to the user, and in this one, an important key is home gateway; basically these are special types of arrangements and groups of connections which enable all the connections to work in a synchronised way and give the maximum output with best results to the user.

9.8 Algorithms and Methods

Algorithms and its methods are useful in providing an interactive and user-friendly environment in the home or smart home. The algorithms help to detect the location accurately and all the activities which are based on location and its usage. Algorithms and its methods are also useful in collecting data and its various attributes to predict the best possible future [33]. Neutral networks are still in demand and are very popular with such complexity of construction like in the smart homes.

9.9 Future Challenges

Future homes are no doubt a good concept, but still there are many hidden and non-hidden challenges which are not enabling the user to use it more effectively. The security issues linked with the smart homes are quite challenging for the makers because nowadays user privacy and security is the topmost priority, and if the smart home is not able to provide this too, then there is no use of calling it as a smart home or the home of the future [34, 35]. The future of these homes will totally depend upon how technology will advance, and its issues are recognised [53]. These smart homes are really good in providing comfort to its users, but its technology needs to be flexible, that is, it should be all user-friendly; only then can we predict the future of the smart homes [36]. These smart homes are generally based on special type of engineering and need special attention, weather in terms of connecting the communications, sensors and information technology.

There are various protocols and special types of adjustment which are required to construct the smart homes so that they could be designed in such a way that in turn will give comfort to the user. The home is also equipped with several medical and fitness applications, but their capability in meeting the requirements of the user and working as per instructions given by the user should be ensured [37]. to upscale the advantages of smart homes, then deeper study needs to be done that what makes do convenient that user will buy it, erase another option of buying: means constructing such markets for these types of smart homes and where no other will be seen.

As we are talking about smart homes and its usage, what about power grid, stations and electricity? What will happen to the applications during electricity breakdown? These are some of the important questions that still need to be answered because all are very relevant questions if one is planning to live in a smart home [38, 39]. The future of smart home can be bright but after removing all the barriers of another option. User-friendliness and comfort are all that a customer wants from such type of smart homes.

9.10 Conclusion

Smart homes and its future in the upcoming scenario seem to be quite easy as many technologies still exist, and many of them are user-friendly. Ensuring the demands of the user that what the user wants, how the user wants, and what type of technology especially user wants in their smart homes, needs to be in the account. Smart homes will cover many opportunities in the upcoming scenario of housing and living [40, 41].

Living is never so much easy if we talk about India, where many of the people are still not addicted or prone to new technology [42, 43]. We can say that people are not technology-friendly here. Smart home is a home consisting of all the needs and wants the people need to maintain their luxury and comfort level. It is all about how you adapt to these technologies and how you change with this. Smart home and its future will totally depend upon how much the population of the specific state is educated, because these types of things cannot be easily bought by the less educated. The higher the education level, the higher will be the standard of living and the higher will be the expenditure on it to buy that living [44].

The smart will surely be user-friendly as we discussed above as well; like in terms of service integration and its usage, various combining of technology and sensors installed in these types of smart homes, but nowadays people are more concerned of the security issues linked with it or which can arise in the future after its usage gets started [45, 46]. Security is one of the key concerns which none of the user wants to compromise, and the user only limits itself from not opting for these smart homes because of the security and technological issues. The company needs to develop some amount of confidence in the user regarding buying of the smart homes and its usage [47]. In the upcoming time, the concept of smart homes will be a much discussed phenomena based on the analysis of recent trend which indicates the centre of intelligent service consumption. As the present generation is more inclined towards luxury and comfort, then no doubt smart homes will be a profitable market in the future, and it will be a great revolution. Christos Stergioua et al. [48] merge cloud computing and IoT to show how the cloud computing technology improves the functionality of the IoT. Smart home and IoT are rich with sensors, which generate massive data flows in the form of messages or events. Processing this data is above the capacity of a human being's capabilities [49]. The level of intelligence or controllability of a smart home service that users want may differ according to the user. As potential users of smart home services have diversified in recent years, providing the appropriate functions and features is critical to the diffusion of the service. Thus, this study examines the smart home service features that current users require and empirically evaluates the relationship between the critical factors and the adoption behaviour [50]. Vadillo et al. [51] conducted research on telecare system adoption, which is one type of smart home service, and found perceived usefulness is important for the intention to use the system.

References

1. http://www.un.org/News/Press/docs//2007/pop952.doc.htm. Last visited 1 Dec 2008.
2. http://www.who.int/disabilities/en/. Last visited 23 June 2009.
3. Chan, M., Estève, D., Escriba, C., & Campo, E. (2008, July). A review of smart homes-present state and future challenges. *Computer Methods and Programs in Biomedicine, 91*, 55–81.
4. Hightower, J., & Borriello, G. (2001, August). Location systems for ubiquitous computing. *Computer, 34*, 57–66.
5. Manley, E. D., Nahas, H. A., & Deogun, J. S. (2006). Localization and tracking in sensor systems. In *Proceedings of the IEEE international conference on sensor networks, ubiquitous, and trustworthy computing* (pp. 237–242).
6. Robles, R. J., & Kim, T.-H. (2010, January). Review: Context aware tools for smart home development. *International Journal of Smart Home, 4*, 1–12.
7. Pishva, D., & Takeda, K. (2006). A Product Based Security Model for Smart Home Appliances. In *Proceedings of 40th annual IEEE international Carnahan conference security technology* (pp. 234–242).
8. Pishva, D., & Takeda, K. (2008, October). Product based security model for smart home appliances. *IEEE Aerospace and Electronic Systems Magazine, 23*, 32–41.
9. Lutolf, R. (1992). Smart home concept and the integration of energy meters into a home based system. In *Proceedings 7th international conference metering apparatus and tariffs for electricity supply* (pp. 277–278).
10. Berlo, A. V., Bob, A., Jan, E., Klaus, F., Maik, H., & Charles, W. (1999). *Design guidelines on smart homes, A COST 219bis Guidebook*. European Commission.
11. Winkler, B. (2002). *An implementation of an ultrasonic indoor tracking system supporting the OSGi architecture of the ICTA lab*. Master's thesis, University of Florida.
12. Briere, D., & Hurley, P. (2003). *Smart homes for dummies*. New York: John Wiley & Sons, Inc.
13. http://www.changeagentteam.org.uk/_library/docs/housing/smarthome.pdf. Last visited 3 Dec 2008.
14. Satpathy, L. (2006). *Smart housing: Technology to aid aging in place. New opportunities and challenges*. M.S. dissertation, Mississippi State University.
15. Das, S. K., Cook, D. J., Battacharya, A., Heierman, E. O., III, & Lin, T.-Y. (2002, December). The role of prediction algorithms in the MavHome smart home architecture. *IEEE Wireless Communications, 9*, 77–84.
16. Youngblood, G. M., Holder, L. B., & Cook, D. J. (2005). Managing adaptive versatile environments. In *Proceedings 3rd IEEE international conference pervasive computing and communications* (pp. 351–360).
17. Gopalratnam, K., & Cook, D. J. (2007, January–February). Online sequential prediction via incremental parsing: The active LeZi algorithm. *IEEE Intelligent Systems, 22*, 52–58.
18. Youngblood, G. M., & Cook, D. J. (2007, July). Data mining for hierarchical model creation. *IEEE Transactions on Systems, Man, and Cybernetics – Part C: Applications and Reviews, 37*, 561–572.

19. Zhang, S., McClean, S., Scotney, B., Hong, X., Nugent, C., & Mulvenna, M. (2008). Decision support for Alzheimer's patients in smart homes. In *Proceedings 21st IEEE international symposium on computer-based medical systems (CBMS)* (pp. 236–241).
20. Mozer, M. C. (1998). The neural network house: An environment that's adapts to its inhabitants. In *Proceedings AAAI spring symposium on intelligent environments* (pp. 110–114).
21. Helal, S., Winkler, B., Lee, C., Kaddoura, Y., Ran, L., Giraldo, C., Kuchibhotla, S., & Mann, W. (2003). Enabling location-aware pervasive computing applications for the elderly. In *Proceedings IEEE 1st conference pervasive computing and communication* (pp. 531–536).
22. Helal, S., Mann, W., El-Zabadani, H., King, J., Kaddoura, Y., & Jansen, E. (2005, March). The gator tech smart house: A programmable pervasive space. *Computer, 38*, 50–60.
23. Noguchi, H., Mori, T., & Sato, T. (2002). Construction of network system and the first step of summarization for human daily action in the sensing room. In *Proceedings IEEE workshop on knowledge media networking (KMN'02)* (pp. 17–22).
24. Brdiczka, O., Crowley, J. L., & Reignier, P. (2009, February). Learning situation models in a smart home. *IEEE Transactions on Systems, Man, and Cybernetics. Part B, Cybernetics, 39*, 56–63.
25. Krumm, J., Harris, S., Meyers, B., Brumitt, B., Hale, M., & Shafer, S. (2000). Multi-camera multi-person tracking for EasyLiving. In *Proceedings 3rd IEEE international workshop on visual surveillance* (pp. 3–10).
26. Brumitt, B., Meyers, B., Krumm, J., Kern, A., & Shafer, S. (2000). EasyLiving: Technologies for intelligent environments. In *Proceedings 2nd international symposium on handheld and ubiquitous computing* (pp. 97–119).
27. Yamazaki, T. (2006). Beyond the smart home. In *Proceedings international conference hybrid information technology (ICHIT'06)* (pp. 350–355).
28. Swaminathan, R., Nischt, M., & Kuhnel, C. (2008). Localization based object recognition for smart home environments. In *Proceedings IEEE international conference multimedia and expo* (pp. 921–924).
29. Moeller, S., Krebber, J., Raake, A., Smeele, P., Rajman, M., Melichar, M., Pallotta, V., Tsakou, G., Kladis, B., Vovos, A., Hoonhout, A., Schuchardt, D., Fakotakis, N., Ganchev, T., & Potamitis, I. (2004). Inspire: Evaluation of a smart-home system for infotainment management and device control. In *Proceedings language resources and evaluation (LREC)* (pp. 1603–1606).
30. Kumar, S., & Qadeer, M. A. (2009). Universal digital device automation and control. In *Proceedings 2nd IEEE international conference on computer science and information technology (ICCSIT)* (pp. 490–494).
31. Alahakoon, D., Halgamuge, S. K., & Srinivasan, B. (2000, May). Dynamic self-organizing maps with controlled growth for knowledge discovery. In *IEEE transactions on neural networks* (Vol. 11, pp. 601–614).
32. Zheng, H., Wang, H., & Black, N. (2008). Human activity detection in smart home environment with self-adaptive neural networks. In *Proceedings IEEE international conference on networking, sensing and control (ICNSC)* (pp. 1505–1510).
33. Shehata, M., Eberlein, A., & Fapojuwo, A. O. (2007). Managing policy interactions in KNX-based smart homes. In *31st annual international computer software and applications conference*.
34. Lu, C.-H., & Fu, L.-C. (2009, October). Robust location-aware activity recognition using wireless sensor network in an attentive home. *IEEE Transactions on Automation Science and Engineering, 6*, 598–609.
35. Ma, T., Kim, Y. D., Ma, Q., Tang, M., & Zhou, W. (2005). Context-aware implementation based on CBR for smart home. In *Proceedings international conference wireless and mobile computing networking and communications (WMOB'2005)* (Vol. 4, pp. 112–115).
36. Chen, C.-Y., Tsoul, Y.-P., Liao, S.-C., & Lin, C.-T. (2009). Implementing the design of smart home and achieving energy conservation. In *Proceedings 7th IEEE international conference on Industrial Informatics (INDIN)* (pp. 273–276).

37. Rashidi, P., & Cook, D. J. (2008). Keeping the intelligent environment resident in the loop. In *Proceedings 4th international conference on intelligent environments* (pp. 1–9).
38. Perumal, T., Ramli, A. R., & Leong, C. Y. (2008, May). Design and implementation of SOAP-based residential management for smart home systems. *IEEE Transactions on Consumer Electronics, 54*, 453–459.
39. Wang, Z., Wei, S., Shi, L., & Liu, Z. (2009). The analysis and implementation of smart home control system. In *Proceedings of international conference information management and engineering* (pp. 546–549).
40. Yongping, J., Zehao, F., & Du, X. (2009). Design and application of wireless sensor network web server based on S3C2410 and Zigbee protocol. In *Proceedings of international conference on networks security, wireless communications and trusted computing* (Vol. 2, pp. 28–31).
41. Wu, C., Liao, C., & Fu, L. (2007). Service-oriented smart-home architecture based on OSGi and mobile-agent technology. *IEEE Transactions on Systems, Man and Cybernetics, Part C: Applications and Reviews, 37*, 193.
42. http://www.knopflerfish.org/. Last visited 25 Apr 2011.
43. Nikolaidis, A. E., Papastefanos, S., Doumenis, G. A., Stassinopoulos, G. I., & Drakos, M. P. K. (2007, October). Local and remote management integration for flexible service provisioning to the home. *IEEE Communications Magazine, 45*, 130–138.
44. Virone, G., Noury, N., & Demongeot, J. (2002, December). A system for automatic measurement of circadian activity deviations in telemedicine. *IEEE Transactions on Biomedical Engineering, 49*(part 1), 1463–1469.
45. Mihailidis, A., Carmichael, B., & Boger, J. (2004, September). The use of computer vision in an intelligent environment to support aging-in-place, safety, and independence in the home. *IEEE Transactions on Information Technology in Biomedicine, 8*, 238–247.
46. Farella, E., Falavigna, M., & Ricco, B. (2009). Aware and smart environments: The Casattenta project. In *Proceedings 3rd international workshop on advances in sensors and interfaces (IWASI)* (pp. 2–6).
47. Adlam, T., Faulker, R., Orpwood, R., Jones, K., Macijauskiene, J., & Budraitiene, A. (2004, September). The installation and support of internationally distributed equipment for people with dementia. *IEEE Transactions on Information Technology in Biomedicine, 8*, 253–257.
48. Mucheol Kim, Ka Lok Man, Nurmamat Helil, "Advanced Internet of Things and Big Data Technology for Smart Human-Care Services", Journal of Sensors, vol. 2019, Article ID 1654013, 3 pages, 2019. https://doi.org/10.1155/2019/1654013
49. Vadillo, L., Martín-Ruiz, M. L., Pau, I., Conde, R., & Valero, M. Á. (2017). A smart telecare system at digital home: Perceived usefulness, satisfaction, and expectations for healthcare professionals. *Journal of Sensors, 2017*, 8972350., 12 pages.
50. Stergioua, C., Psannis, K. E., Kimb, B.-G., & Gupta, B. (2018, January). Secure integration of IoT and cloud computing. *Elsevier, Future Generation Computer Systems, 78*(Part 3), 964–975.
51. Khan, N. S., Ghani, S., & Haider, S. (2018). Realtime analysis of a sensor's data for automated decision making in an IoT-based smart home. *Sensors*. https://doi.org/10.3390/s18061711.
52. Kumar, A., Kumar, P. S., & Agarwal, R. (2019). A face recognition method in the IoT for security appliances in smart homes, offices and cities. In *2019 3rd international conference on computing methodologies and communication (ICCMC)*. IEEE.
53. Garg, V., & Agrawal, R. (Eds.). (2020). *Transforming management using artificial intelligence techniques*. CRC Press.

Chapter 10
Providing Security and Managing Quality Through Machine Learning Techniques for an Image Processing Model in the Industrial Internet of Things

B. Vineetha and R. B. Madhumala

10.1 Introduction

The reliance on technology has growing enormously; however, there are a few domains where the connection with Information Technology (IT) is essential. There is a newfound technology dependence which has led to huge modifications in the way we interact with the environment. Devices are being connected all around and used for huge areas. Nowadays, most devices have inbuilt sensors which collect information and enable instruments to interrelate with one another. Every piece of data gathered will later be sent to humans affecting the demand of devices. The system for managing quality depends on industrial image processing that permits devices to interlink through the cyber physical cloud system (CSPS). Some items are typically produced most in industries, and mass production leads to various challenges, such as price, efficiency, time consumption and quality issues. To control these issues, most of the industries have shifted to automation. The excellence of the products is maintained until the last item of the manufacturing line to ensure quality throughout the line and check the last item for faults. The majority of manufacturing companies employ computerized structures that use image processing to determine every problem. The image processing structures act as a very important function in the automated industrial production processes. The accessibility of extremely competent recent interfaces for communication along with cameras enables faster image processing. The image processing software quality has becomes extremely dependable even as concurrently dropping the cost. The two important criteria—central processing unit (CPU) and time efficiency—are improving every day. The highest priority among manufacturers is providing quality goods.

B. Vineetha (✉) · R. B. Madhumala
Department of Computer Science and Engineering, PES University, Bengaluru, India

Department of Computer Science and Engineering, JAIN University, Bengaluru, India

M. Moh et al. (eds.), *Smart IoT for Research and Industry*, EAI/Springer Innovations in Communication and Computing,
https://doi.org/10.1007/978-3-030-71485-7_10

The machine can implement powerful computer vision which has to be built such that the parts can be identified throughout the manufacturing process. The research in the machine vision models in manufacturing companies has achieved superior potential along with power similar to the current machine learning structure. Research mainly concentrates on making vision systems more competent and cost effective. Currently, based on the classification, the methods or techniques used during inspection may change. The machine vision system research investigates the device structural quality that has been completed in various domains. Other research focuses on the diverse techniques and methods for computerized visual examination of printed circuit boards (PCBs). The automated inspection systems use industrial image processing. The image quality will be increased only to improve features implemented for the image processing operations. The image analysis will be increased; the image analysis allows pattern recognition. It mainly focuses on result commonality along with image patterns which helps in differentiating the things available in the image. The machine learning with pattern identification will break down an image to productively notice faults in a structural thing. The apparatus used for analysis depends on different machine learning algorithms. Visual inspection systems dependent on the machine learning classically employ supervised learning algorithms with support vector machines (SVMs). In order to identify faults, multi instance learning is used. Machine intelligence learning has been previously used for image retrieval, object recognition, target tracking and image categorization. The main focus has been on quality management throughout CPUs manufacturing. The many components could be missing or misplaced causing errors throughout assembly of the CPU. In order to avoid this, there should be some protocols which guarantee the product quality. A system has to be developed to certify quality, efficiency and lower time costs at the final stage of the production line. There are diverse components in CPU, along with all components, which have to be investigated for faults. The machine learning algorithms are capable of being utilized to resolve the problem. There should be some competent approach to all devices engaged in the interaction with each other. Sensors, actuators and other devices interact with each other and are interlinked to form Smart Networked Systems and Societies (SNSS). There are no limitations to the fixed components, and they can comprise mobile devices also. The bridge between the virtual and physical world will be formed through cyber physical systems by SNSS. The purpose of cyber physical systems is to link the cyber world and humans via sensors and actuators. The cyber physical systems link myriad devices and results in a huge deep data swap. The cyber physical systems will manage the various resources connected with the system. It includes data services along with components such as sensors, actuators and processors. The cloud servers are used to perform different operations. The bandwidth, energy and storage will be dispersed by the cloud servers based on their demands. A cyber physical system is considered as a difficult system which controls or monitors and exchanges data in the real world. The complete infrastructure will be well operated. Because the device capacity and organizations are attached to this, a small error will lead to extra repercussions on the entire scheme. It helps industry manage the difficult industrial processes through sensors, actuators and processors.

The Industrial Internet of Things was initially presented as an idea by the Government of Germany and will be the fourth Industrial revolution. Industry 4.0 is the combination of automation [5], Internet of Services [6], augmented reality [7], wireless technologies along with concentric computing [8], cyber physical systems [1], Internet of Things [2–4] and cloud computing. More associated areas such as the IoT, Big Data Analytics and cloud computing will enable Industry 4.0 to obtain unmatched flexibility, exactness and competence in developed methods [9, 10]. Industry 4.0 aims to achieve certain interoperability, orientation of service, modularity and security from end-to-end using cross-platform combinations [11], and there are a few merits such as self-awareness, self-prediction and self-organization for a system [12]. The allocation value for captured data from cross-platform integration will occur using big data analytics within Industry 4.0. Big data analytics refers to the gathering, controlling, processing and analysing of data while endlessly ensuring the provisions of volume, velocity, value, variety and veracity [13]. The unobstructed interior and exterior actions are suitable for customers using Industry 4.0 big data [14]. Every data collected will be managed by the system using more impermanent and constant storage that allows in-memory, on-board and a huge range of scattered storage activities across Industry 4.0 structures [15, 16]. The data granularity processing services for big data analytics processes in Industry 4.0 systems are diverse, starting with IoT instruments resource-constrained to the ingenious huge-scale spread of the cloud computing systems [17], and there are differences in descriptive, prescriptive and preventative actions for analytic operations [14].

10.2 Background

The physical evaluation of all the products is time consuming and pricey, as well as sometimes inaccurate and complicated. The quality management and control in several industries helps to overcome this by supporting automated services. A huge part of automation is image processing. A variety of methods are able to computerize the management of quality in a manufacturing company. The defects in the items are determined using image processing. The finding of faults in items and misplaced parts in items has to be identified at the beginning. Some studies have been carried out with the aim of finding an equation to compute the outline of mislaid areas. The missing areas in each image were separately determined approximately by combining a collection of images to secure the entire translation of the original image.

Image pre-processing involves evaluating the value of the item by taking the images of the items to be pre-processed. The most important step is where the image captured is altered to improve the image data. Machine Intelligence (MI) learning is the main method for outline recognition along with computer vision. This is a section of machine learning. There are object ambiguities which are classified in visual inspection applications. Let's consider the bags which comprise multi-sets instances. Every instance group is characterized via an attribute vector along with restrictions to communicate each bag label to unknown labels in every instance. Both MI learning and image processing methods are used to recognize faults in a CPU production line. Cyber physical system has cost-effective aspects and is a system in which a massive number of devices will be connected. It has high power expenses because of its Information and Communications Technology (ICT) requests. A constraints group along with requirements to achieve an energy-efficient cyber physical system is not merely about the energy allocation but also concerns resources and services. Demand-side management (DSM) is necessary to exact resources use at a manufacturing company. The resource demand will differ and is sometimes uncontrollable. Maintaining an efficient DSM is a difficult task because much variation may occur. When considering only the industrial image processing feature, a few issues are as follows:

- Demand in daily deviation.
- Demand in seasonal deviation.
- There are interruptions which are inclined to be more costly.

Low delay and high throughput are the significant parameters to think about in any system. These parameters encompass a straight effect on the time occupied for any kind of message. Various investigations have been done to determine the organization of power in cyber physical systems. A distributed system is the main characteristic of a cyber physical system which is distributed to all so that demand and power supply is a big issue. The data centres manage policy anywhere the energy expenses are deduced while at the same time increasing the quality of computational services. The association of network connectivity and physical objects which gather and swap data is referred to as the Internet of Things (IoT). A device which links to the internet and sends the device data to other devices is referred to as a 'thing'. The main requirement of the IoT is IPv6 Internet which introduces challenges to append billions of devices to the IPv4 Internet [23, 24].

10.3 IoT Security

From the beginning of the twenty-first century, the safekeeping of IoT devices has shown to be a difficult task. The IoT connects distant devices and links the whole world, also opening a diversity of windows that can be exploited by a variety of attack types [37]. The phrase the IoT encompasses the whole globe by means of smart methods and functions, some of which can be fictional. In 1999, Kevin Ashton [11] used IoT for the first time.

This use suggested a connection among the virtual world and human beings by means of various smart devices through their functions using different communication rules. In the present, the higher altitude of the globe is covered with smart technology; moreover, the IoT is the major component similar to the heart of it. Currently, people cannot consider a solitary instant without applying IoT devices and their functions. Almost 50 billion instruments will be linked with the Internet by

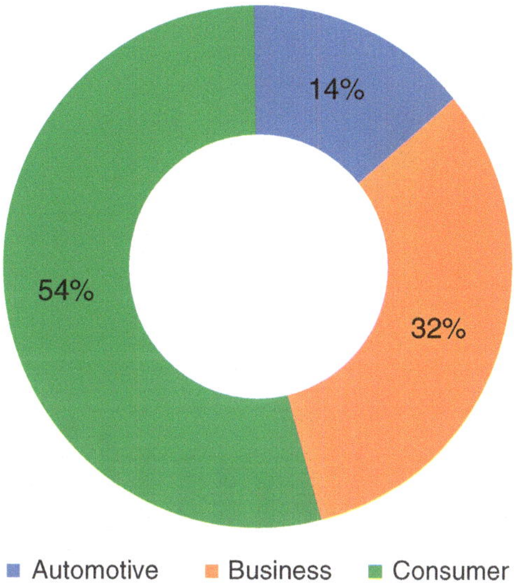

Fig. 10.1 Approximate IoT device users by 2020

2020, as indicated in a review. This will increase exponentially as time passes [12]. The approximate number of IoT devices used by 2020 is shown in Fig. 10.1 [13].

This will also include approximately USD 3.9–11.1 trillion inexpensive markets by 2025 [14]. In addition to the global market of the IoT system, the number of IoT devices connected currently and the prediction of future devices by 2025 [15, 16] is given in Fig. 10.2. The study of the IoT plus its progress along with defence has recently received extensive attention in the areas of Computer Science and Electrical Engineering.

10.4 IoT Layers

The IoT architecture is a doorway of diverse uses, and it is built to set up an association and enlarge IoT functions at each doorway. The diverse rules communiqué together with Wi-Fi, Bluetooth, RFID, ZigBee and LPWAN are incorporated in dissimilar IoT structural design layers to send and accept varied information [17, 18, 37]. In addition, huge scale advanced companies have their own platforms of IoT to service their expensive clients [19]. A paradigm structural design of the IoT encompasses primarily three layers, as shown in Fig. 10.3 [20]:

- Physical layer or Perception layer
- Network layer
- Application layer or Web layer

The Perception layer is the last layer in the IoT architecture which provides functions to the clients using mobile and web-based software. Based on innovative scope

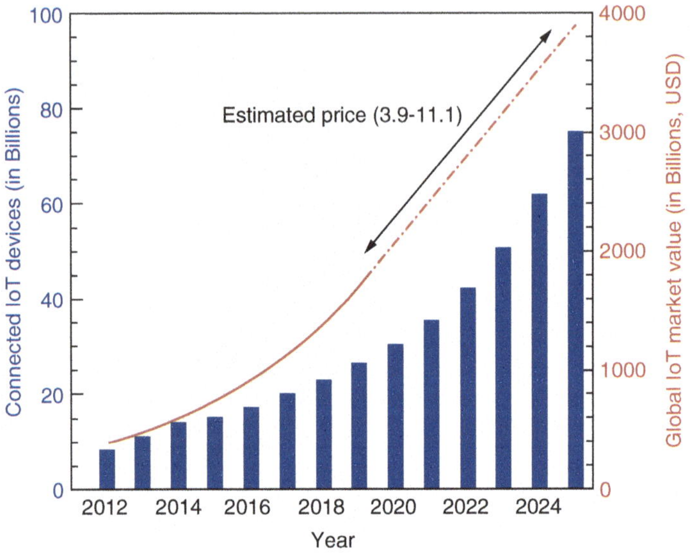

Fig. 10.2 Graphical arrangement of all linked IoT devices as well as global IoT market so far and future calculation

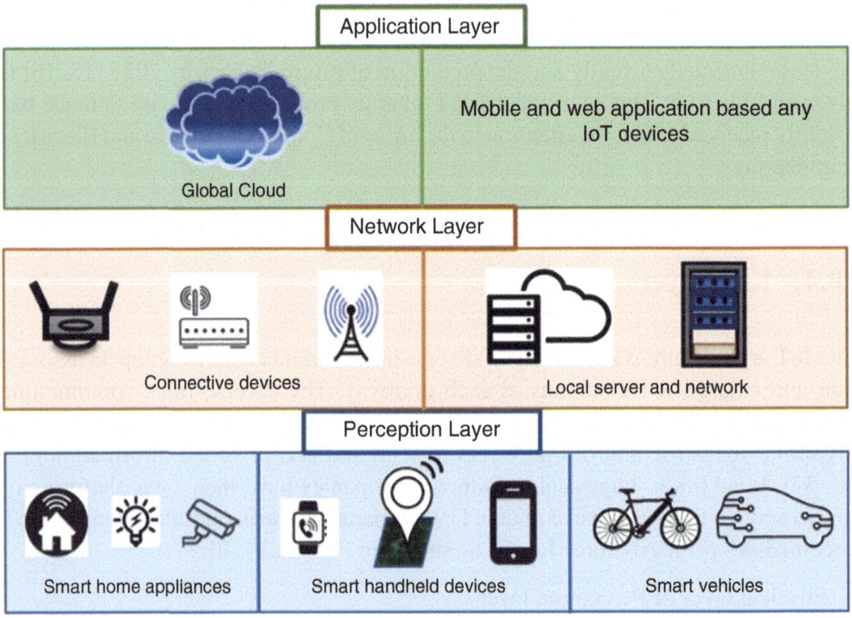

Fig. 10.3 IoT architecture

and applications of smart things, there are many uses for the IoT in this technologically advanced world. For example, smart devices and services connecting to the IoT have been developed for health, education and transportation [21, 22].

In IoT systems, the Network layer acts as a transfer and receiving middle layer for information using a variety of connection rules that join devices with smart services [23]. Some servers are local clouds that store as well as process the data which function as a middleware with the system and the following layer [24–26]. The extra significant feature is big data in this layer due to its properties regarding the currently ever-growing cheap market. The items from the physical layer are generating a vast amount of information frequently that is being sent, processed and stored by IoT systems. As information are vital in the network layer, machine learning and deep learning for smart functions are primarily used at the instant to examine the data stored to utilize improved analysis techniques and carry out high-quality use from it for smart devices [27].

The Perception layer is the IoT architecture's first layer that encompasses the Physical followed by Medium Access Control (MAC) layers. The physical layer mainly deals with hardware which is used to transmit and accept information using a variety of communication rules [28–31]. The establishment of a connection links physical devices with networks to allow the appropriate communication through the MAC layer. MAC uses diverse rules to connect with network layers. Most instruments in IoT layers are the plug and play kind where a vast part of big data is created [32–36].

10.5 Proposed Research

The items are produced in a batch in an industry. The amount of items to be processed for quality control could be overwhelming for the industry. A small flaw in the CPU production line such as a misplaced screw might go unnoticed when being hand processed. The purpose of the proposed system is to find abnormalities in the items. The images of end products are taken at the final stage of the production lines of the systems. The images are used to help to identify faults in items.

The images are captured by the cameras located at the end of the production line which includes sensors. The cameras are located in such a way that the whole batch of things can be investigated using the images taken by the cameras. It is important for the items to have been imaged from each angle. Image processing in industries is used to check the items that are produced, and images are captured throughout the production line that then have to be pre-processed. The images pre-processed have to be transmitted in a method which is split and includes the angles of the item.

1. The system comprises four modules, as shown in Fig. 10.4.
2. To separate the images into different components use edge detection.
3. The different parts to discover lost parts to be processed.
4. The description of missing part is to be predicted.

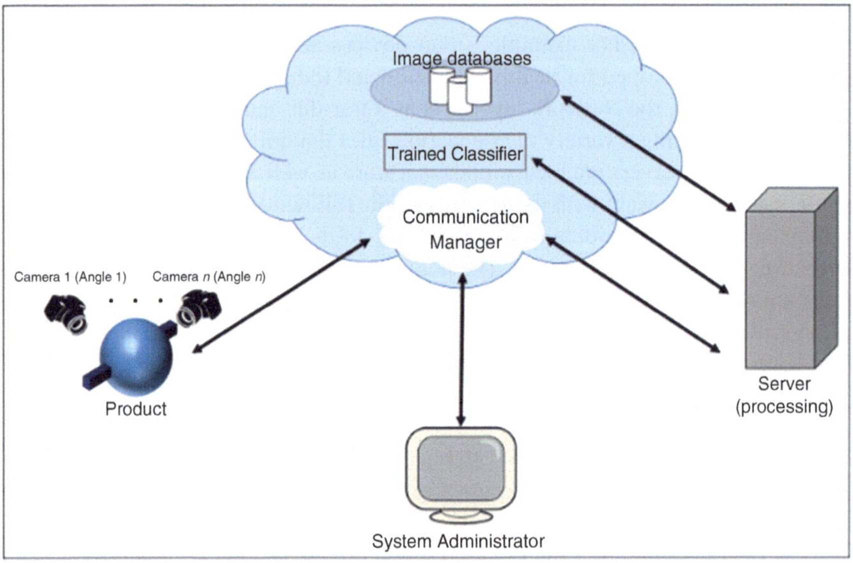

Fig. 10.4 Various modules of representation

5. Camera occlusion to be corrected.

Edge detection is used to separate the image into different parts. The image is to be scanned first and split into diverse parts. The stages of primary edge detection are:

1. The smoothing process helps to suppress the noise.
2. The edge enrichment step sorts the image for contrast.
3. The purpose of edge localization is that the local maxima from the sort outcome are really edges versus noise determined using thresholding. The threshold value is particular. Pixels less than the thresholds are set to 0, pixels greater than the thresholds are set to 1.

The stages to image a segment into the variety of parts using canny edge detector includes:

1. The image is sorted with a Gaussian derivative.
2. Check out the degree and point of reference of the gradient.
3. Non-maximum suppression is used to reduce extensive ridges down to solitary pixel size.
4. Linking and thresholding are carried out to explain the low and high threshold. The high threshold is used to start edge curves and the low threshold is used to preserve them.
5. The image is segmented into a range of components.

The collective with thing recognition outcome forecasts where things are lost in an image using convolution neural network structure. The procedure disregards things and concentrates on framework only. For example, consider the lost parts in a production line. A thing can be clear as lost in an image area while:

1. The search for the thing finds nothing.
2. The forecast of the thing's characteristic surroundings elevates the probability of its subsistence.

The aim is to find all such regions efficiently in a given image. The need is to instruct this demonstration to place lost things. The intention of the module is to scan an image and discover the lost things. Because the image is before now separated into parts, this stage builds a mechanized algorithm to scrutinize the parts to make out if any part is lost. The algorithms specifically enable parts of the production line to maintain and enhance the capability as well as subsequently improve the proceeds of industry. The algorithm has to account for all parts, including the lost parts. To create a representation, this concentrates on the structure and consequently further behaves successfully in the proposed system. The framework is an assembly line in CPU. Therefore, the representation will have to be trained using images in this region that existed before. The benefit of this is that the representation will have to look for correct images of standard parts and this also simplifies training. The successful fault finding instructs the system using the machine learning technique. The feature space F is also called instance space. The training dataset $T = \{(B1, l1), (B2, l2), \ldots, (Bn, ln)\}$, where$\{B1, \ldots, Bn\}$ $T = \{(B1, l1), (B2, l2), \ldots, (Bn, ln)\}$, where$\{B1, \ldots, Bn\}$ is a bag set and each $BiBi$ is a multiset of ni instances; $Bi = \{Bi, 1, \ldots, B1, ni\}$ and $L = \{l1, l2, \ldots, ln\}$, $li \ \varepsilon\{-1, +1\}ni$ instances; $Bi = \{Bi, 1, \ldots, B1, ni\}$ and $L = \{l1, l2, \ldots, ln\}$, $li \ \varepsilon\{-1, +1\}$ are the labels of class. The supposition is that a bag represents a thing that has to be observed for a fault and the label refers to whether a characteristic in the feature space is there or not. In this case of a visual inspection system, every instance has to be positive for a bag to be characterized as positive. The intention of this categorization scheme is to anticipate the tag of an unnoticed bag

$$f(B): B -> L$$

A bag represents an image and all the likely faults are the instances. Therefore, if a solitary instance is positive, it means that a fault is there. MI learning classifier takes the constitution

$$\boldsymbol{f(Bi) = \{+1,|,if\, 8 \neg 3 Bi,|,j \varepsilon Bi,|,f(Bi,j) = +1 - 1,|,}\text{otherwise}\}$$

The machine learning classification algorithm is:

Input: The training set $\{(Ii, li) | 1 < = i < = n\}$where Ii is an image set$\{(Ii, li) | 1 < = i < = n\}$where Ii is an image set.

Output: A classifier trained. $f(B)f(B)$

1. Generate a blank training set, $T < - \{\} T < - \{\}$
2. For all images $IiIi$

3. Generate a null bag, $Bi < -\{\} Bi < -\{\}$
4. Retrieve all available faults in $EiEi$
5. For all defects $ei, j\varepsilon Eiei, j\varepsilon Ei$
6. Retrieve instances into $B i, jB i, j$
7. Include the recent instance to the bag $Bi = BiUBi, jBi = BiUBi, j$
8. Include the latest set, $T < -TU(Bi, l i)T < -TU(Bi, l i)$
9. $f(B)f(B)$ The trained machine learning classifier of the training set TT

The different parts of the image made available through the algorithms of edge detection are the obtainable faults. The obtainable error is positive, indicates the defective parts and signifies the things' faults. The predictions of lost parts are detailed. The painting region is the technique of reconstructing the deteriorated parts of videos and images. The process can enable us to explain the missing parts in this point. The camera occlusion potential is to be corrected. The object identification is important because the image should have clarity and the complete object should be clearly visible. The situations are not satisfied when they are not able to realize some of the previous modules.

Most challenges still exist regarding the discovery of a thing such as delusion, less visibility, cast shadows and occlusion of object. When the sight of the camera is somewhat blocked, it is called camera occlusion. Self-occlusion and inter-object occlusion are the two categories of occlusion. The purpose of this element is to control the cameras so as to confine images of the connected items in such a way that the item is not occluded. The proposed solution handles occlusion in different areas such as to overcome vehicle collision in urban crowded roads. Here the camera is adjusted to obtain images of the entire thing. The total image cannot be incorporated in a solo image; therefore, a supplementary image can be built to utilize and observe a fault in an item. The occlusion handling helps in the arrangement to choose if the created images by a camera can be dependable or any images in addition to the solitary images are central to construct the essential outcome. The successful performance of all units will require different algorithms.

The image processing system setup comprises the following:

The system sensors for visual quality management are cameras.

The hardware requirement involves use of a corresponding computer because it is an extremely data-intensive procedure. A quick estimation speed and an extremely elevated bandwidth are necessary to hold the data rates of those uses. Apart from the group of developed test, tasks can simply be handled with PCs and hypothesis parts.

The image processing system has to attempt to process regions by parts with the industrialized procedure in addition to being competent to broadcast the outcome such that this whole process can be automated. Subsequently, there has to be a scheme of announcement with earlier instruments. The data necessary is stored in the cloud and the announcement will be accepted through the cyber-physical system network. There is a range of security problems that need to be handled cautiously and economically using the cloud.

10.6 Algorithm

Input: A scheme contains trained data, and the image that has to be processed is shown in Fig. 10.5.

Output: The calibration of the camera if an item has a notice or not.

Assumption: The camera is mechanically calibrated by the scheme.

1. The image is fed into the arrangement and analysed for occlusion.
2. If the image passes the occlusion examination, then the subsequent steps are performed. Otherwise, the recalibration of the camera and a recent image is fed into the scheme.
3. The image segmentation is performed by means of Fuzzy image segmentation.
4. The segmented image splits the image into various components.
5. The categorization of dissimilar parts using machine learning classification.
6. If all the parts are easy to get to, then carry out step 7. Otherwise, carry out step 9.

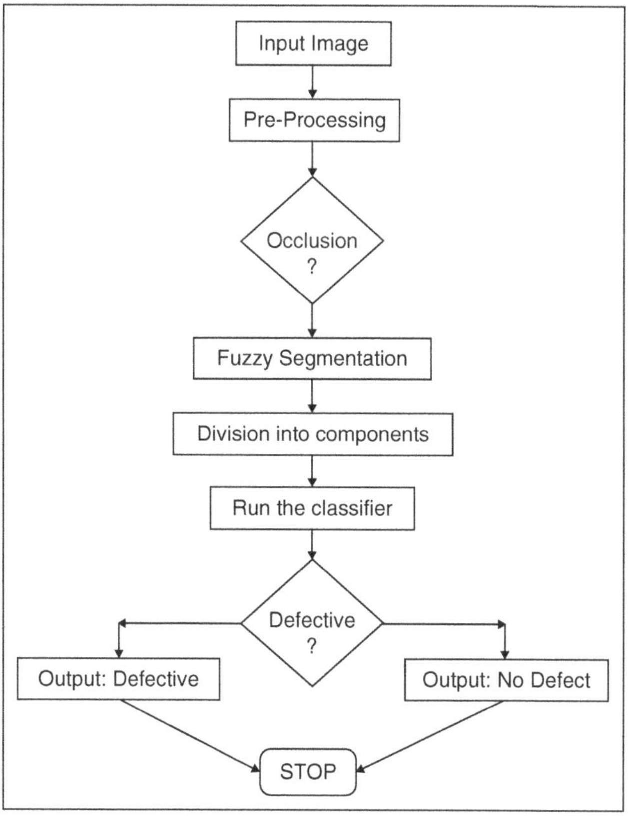

Fig. 10.5 Modules flow chart

7. The parts place in the item is checked.
8. If the locations of the parts are correct, then carry out step 11. Otherwise, step 9.
9. The lost part description and its place are noticed.
10. Defective product as output.
11. Functioning product as output.

10.7 AES Algorithm

The dataset obtained with sensors and used in the machine learning can be secured using a cryptographic algorithm such as Advanced Encryption Standards (AES), also known by its unique name Rijndael, the electronic data encryption recognized by the US National Institute of Standards and Technology (NIST) in 2001. This algorithm includes AES-128, AES-192 and AES-256 block ciphers. The 128-bit key length is used to encrypt as well as decrypt a message block in AES-128. The 192-bit key length is used to encrypt and decrypt message block in AES-192. The 256-bit key length is used to encrypt and decrypt message block in AES-256. Every cipher encrypts and decrypts data in blocks of 128 bits using keys of 128, 192 and 256, correspondingly.

The process of AES mainly involves 4 steps

- AddRound Key
- Substitution and inverse substitution
- Shift rows and inverse shift rows
- Mixcolumn and inverse mixcolumn

AddRound key is the step which is performed in all 10 rounds of AES; before data is given to the substitution step, AddRound key is performed. This involves performing XOR on each bit of plaintext with each bit of sub-key generated by key scheduling algorithm. The output of AddRound key, which is also a state of 128 bits, will be given as input in the substitution step.

The substitution step consists of a substitution table which has an entry of 16×16. Each byte value of state is divided into two parts having 4 bits; the starting 4 bits constitute the row value and the next 4 bits forms the column value. A value of 8 bits is outputted using row and column value as input to the s-box.

The substitution step will produce an output of 128-bit state. This is given as input to the next step which is a shift row function. The shift row function shifts the 2nd row of state by 1 byte, 3rd row by 2 bytes and 4th row by 3 bytes. This is not a binary bit shift, but it is a whole byte shift. The output of AES shift row is also a state of 128 bits, and this becomes input to the last step of

AES round, that is, the mixcolumn. The mixcolumn step is performed in almost all rounds, except in the last round in AES 128 the 10th round is missing the mix-column step. The final output after mixcolumn is given to AddRound key function for the second round, and after 10 rounds, the 128-bit state is generated which has

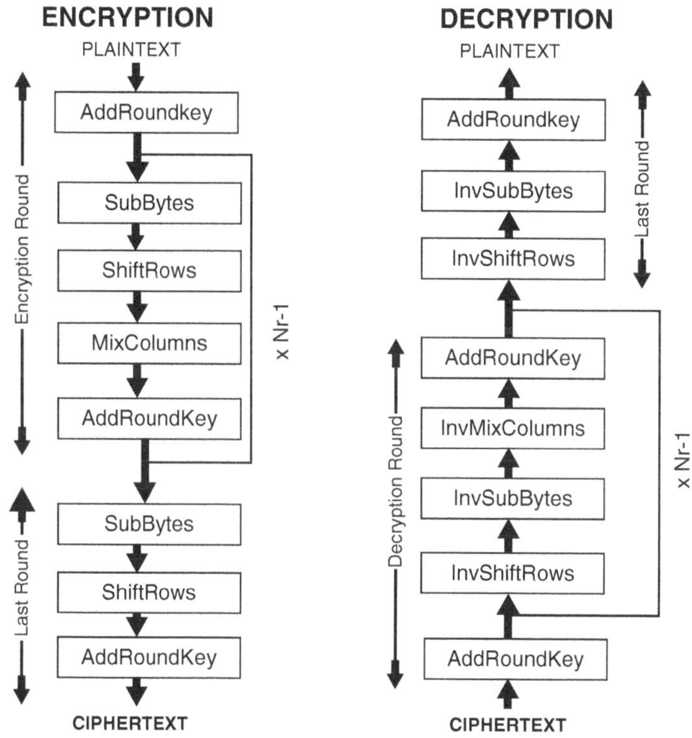

Fig. 10.6 AES encryption and decryption

an avalanche effect on 128-bit input data. In the receiver side, all steps are done in the reverse process.

In AES, all function are invertible functions, that is, $f-1\ (f(x)) = x$ (Fig. 10.6).

10.8 Testing and Result Analysis

The proposed system inspection along with different experiments is performed. The proposed system outcome with preceding existing systems is analysed. The operation uses OpenCV in steps to measure the proposed system, and the primary formed dataset of the image consists of items in the assembly line. Then categorization is performed using four standard classifiers and the proposed system result are analysed. There are 50 defective images and 100 non-defective images in the database. The possible faults are extracted previous to using fuzzy image segmentation as a pattern recognition technique. The dataset has 150 bags (50 positive and 100 negative). There are 271 positive instances and 253 negative instances. In Tables 10.1, 10.2, and 10.3, four state-of-the-art MI learning algorithms are compared with the proposed MI learning algorithm. The classifiers that are being evaluated are

Table 10.1 The proportion of the accurate categorization while finding the flaws in screws

Classifier	Run 1	Run 2	Run 3	Run 4	Run 5
Proposed approach	92.4	93.1	94.2	94.1	93.9
mi-SVM average	90.5	90.4	90.2	91.2	99.6
MIL-Boost	93.2	91.2	90.7	90.9	87.6
MILES	83.5	83.7	79.2	77.6	78.7
MlnD Mean-Mean	90.2	89.3	85.7	82.1	83.9

SVM support vector machine, *MIL* multiple instance learning, *MILES* multiple-instance learning via embedded instance selection.

Table 10.2 The proportion of accurate categorization when there are flaws in labels

Classifier	Run 1	Run 2	Run 3	Run 4	Run 5
Proposed approach	90.7	91.4	92.3	93.1	92.8
mi-SVM average	78.3	79.7	76.4	77.6	79.2
MIL-Boost	83.2	84.6	81.4	81.9	82.0
MILES	69.4	67.5	70.1	70.8	71.2
MlnD Mean-Mean	83.4	83.6	83.7	83.9	84.3

SVM support vector machine, *MIL* multiple instance learning, *MILES* multiple-instance learning via embedded instance selection

obtainable in the multiple instances learning (MIL) MATLAB toolbox. The simulation has been run for a period of time to calculate the correctness of our model with esteem to standard classifiers. The first 5 simulation runs have been displayed in table form to show the accuracy of the proposed model plus preceding standard classifiers. Table 10.1 shows the proportion of the accurate categorization while finding the flaws in screws. Table 10.2 shows the proportion of accurate categorization when there are flaws in labels. Table 10.3 shows the proportion of accurate categorization when there are flaws in sockets.

The proposed representation has the capability to observe whether the screw has flaws, is lost or slackly screwed. Some recreation runs reinforce that each classifiers has been examined keenly. Every classifier has different finding capabilities in varied situations. From the outcome, we can clearly see that the mi-SVM average classifier has the complicated accuracy in judging faulty screws or else sockets excluding lower capacity in deciding faulty labels. Similarly, preceding paradigm classifiers also include an ability to discover lost, lose or faulty screws and sockets except that not all systems are reasonably high-quality at label discovery, but the proposed form is in reality well-organized in discovering all kinds of faults such as missing, drooping or defective parts of the production. By keenly observing on top of tables, it can be illustrated that the assessment on labels was mostly well informed. However, a variety of algorithms do part work a different way for dissimilar kinds of errors. One main cause is the ways the various segmentation algorithms use for straightening out the image into a variety of parts.

Table 10.3 The proportion of accurate categorization when there are flaws in sockets

Classifier	Run 1	Run 2	Run 3	Run 4	Run 5
Proposed approach	92.3	92.9	93.7	94.0	92.8
mi-SVM average	91.9	91.6	90.8	91.7	91.3
MIL-Boost	93.1	92.9	93.4	94.5	93.9
MILES	88.7	88.4	89.5	89.8	87.9
MlnD Mean-Mean	93.4	93.6	92.5	93.1	92.9

SVM support vector machine, *MIL* multiple instance learning, *MILES* multiple-instance learning via embedded instance selection

10.9 Conclusion

The proposed scheme automates the quality estimation of the items formed in an industry. This is achieved by implementing an industrial image processing-based system in a production line using a CPCS. The image processing system examines images of the items in the assembly line for defects. The outcome of this is later communicated to the discerning authority. The main test in our scheme is extracting the instances character. If suitable characteristic extraction methods are not used, it jeopardizes the whole visual inspection scheme. For the prospect to work, we need to imagine the security of the organization. Linking to the cloud throughout communication increases the security challenges. Hence, many methods have to be engaged to attest that the whole CPCS scheme is secure from unnecessary third parties. As a result of IoT growth, any person is able to use and join up their information in the network from wherever using the smart functions of the IoT. Even though, the IoT links lives with the virtual globe through devices to make life easier, safety is a huge concern in the IoT scheme regarding its functions. Therefore, to obtain improved security with time and maintain a high regard, challenges and security of the IoT must be resolved by studying this field and devising solutions and tactics to counter security attacks in the future. A state of the art comprehensive literature review on ML-based security of IoT is obtainable that includes the structural design of the IoT, a comprehensive study on various kinds of safety attacks, attack surfaces with effect, a diversity of categories of ML-based algorithms and ML-based security solutions. In 2018, there was a huge increase in attention towards studies on IoT security. This literature review contains information regarding ML embedded algorithms on IoT security concerning different possible IoT attacks as well as their outside effects. In addition, ML algorithms have been discussed with credible challenges that can aid outlook researchers to update their previous goals and achieve their propose in this field. Increased security of the ML dataset can be provided using lightweight cryptographic algorithms in the future.

References

1. Demant, C., Streicher-Abel, B., & GanicaCarsten, G. (2013). *Industrial image processing: Visual quality control in manufacturing* (2nd ed.). New York: Springer Science + Business Media.
2. Malamas, E. N., Petrakis, E. G., Zervakis, M., et al. (2003). A survey on industrial vision systems, applications and tools. *Image and Vision Computing, 21*(2), 171–188.
3. Moganti, M., Ercal, F., Dagli, C. H., et al. (1996). Automatic PCB inspection algorithms: A survey. *Computer Vision and Image Understanding, 63*(2), 287–313.
4. Kim, K. H., Kim, Y. W., & Suh, S. W. (1998, 13–16 Sept). Automatic visual inspections system to detect wrongly attached components. In *International conference on signal processing applications and technology (ICSPAT'98), Toronto, ON, Canada*. ACM SIGPLAN.
5. Chung, Y., Kim, K. H., & Kim, H. B. (1997, 14–17 Sept). Extracting of line features for automated inspection of vehicle doors. In *International conference on signal processing applications and technology*. San Diego, CA.
6. Paul, A. (2013). High performance adaptive deblocking filter for H.264/AVC. *IETE Technical Review, 30*(2), 157–161.
7. Paul, A., Bharanitharan, K., & Wang, J. F. (2010). Region similarity based edge detection for motion estimation in H.264/ AVC. *IEICE Electronics Express, 7*(2), 47–52.
8. Khandogin, I., Kummert, A., & Maiwald, D. (1998, 13–16 Sept). DSP algorithms for the automatic inspection of fixing devices of railroad lines. In *International conference on signal processing applications and technology (ICSPAT'98), Toronto, ON, Canada*. ACM SIGPLAN.
9. Khalid, W. (1995). A multiscale stochastic image model for automated inspection. *IEEE Transaction on Image Processing, 4*(12), 1641–1654.
10. Paul, A. (2015). Adaptive search window for high efficiency video coding. *Journal of Signal Processing Systems, 79*(3), 257–262.
11. Anand, P., Bharanitharan, K., & Wu, J. J. (2013). Algorithm and architecture for adaptive motion estimation in video processing. *IETE Technical Review, 30*(1), 24–30.
12. Chen, S., Lin, B., Han, X., et al. (2013). Automated inspection of engineering ceramic grinding surface damage based on image recognition. *The International Journal of Advanced Manufacturing Technology, 66*(1–4), 431–443.
13. Unay, D., Gosselin, B., Kleynen, O., et al. (2011). Automatic grading of Bi-colored apples by multispectral machine vision. *Computers and Electronics in Agriculture, 75*(1), 204–212.
14. Yuen, C. W. M., Wong, W. K., Qian, S. Q., et al. (2009). A hybrid model using genetic algorithm and neural network for classifying garment defects. *Expert Systems with Applications, 36*(2), 2037–2047.
15. Paul, A., Jiang, Y. C., & Wang, J. F. (2010). Computation aware scheme for visual signal processing. *JSW, 5*(6), 573–578.
16. Dietterich, T. G., Lathrop, R. H., & Lozano-Pérez, T. (1997). Solving the multiple instance problem with axis-parallel rectangles. *Artificial Intelligence, 89*(1), 31–71.
17. Zhang, C., Chen, X., Chen, M., et al. (2005, 6 July). A multiple instance learning approach for content based image retrieval using one-class support vector machine. In *2005 IEEE international conference on multimedia and expo (ICME 2005), Amsterdam* (pp. 1142–1145). New York: IEEE.
18. Paul, A., Rho, S., & Bharnitharan, K. (2014). Interactive scheduling for mobile multimedia service in M2M environment. *Multimedia Tools and Applications, 71*(1), 235–246.
19. Babenko, B., Yang, M. H., & Belongie, S. (2009, 20–25 June). Visual tracking with online multiple instance learning. In *2009 IEEE conference on computer vision and pattern recognition (CVPR 2009), Miami, FL* (pp. 983–990). New York: IEEE.
20. Paul, A., Pinjari, H., Hong, W. H., et al. (2018). Fog computing-based IoT for health monitoring system. *Journal of Sensors, 2018*, 1386470.

21. Simmon, E., Kim, K. S., Subrahmanian, E., et al. *A vision of cyber-physical cloud computing for smart networked systems*. Gaithersburg, Maryland: US Department of Commerce, National Institute of Standards and Technology.
22. Wang, Y., Vuran, M. C., & Goddard, S. (2008). Cyber-physical systems in industrial process control. *ACM SIGBED Review, 5*, 12.
23. Zhou, J., Kwan, C., & Ayhan, B. (2012, 27–29 Dec). A high performance missing pixel reconstruction algorithm for hyperspectral images. In *2nd international conference on applied and theoretical information systems research, Taipei, Taiwan*. ATISR.
24. Vasilic, S., & Hocenski, Z. (2006, 9–13 July). The edge detecting methods in ceramic tiles defects detection. In *2006 IEEE international symposium on industrial electronics, Montreal, QC, Canada* (pp. 469–472). New York: IEEE.
25. Zhong, F., He, S., & Yi, J. (2015). A fast template matching method for LED chip localization. In *MATEC web of conferences*. EDP Sciences. https://doi.org/10.1051/matecconf/20153404002.
26. Paul, A., Chen, B. W., Bharanitharan, K., et al. (2013). Video search and indexing with reinforcement agent for interactive multimedia services. *ACM Transactions on Embedded Computing Systems, 12*(2), 25.
27. Wu, J., Bai, X., Loog, M., et al. (2017). Editorial of the special issue on multi-instance learning in pattern recognition and vision. *Pattern Recognition*, 444–445.
28. Ray, S., Scott, S., & Blockeel, H. (2011). Multi-instance learning. In C. Sammut & G. I. Webb (Eds.), *Encyclopedia of machine learning* (pp. 701–710). New York: Springer US.
29. Reka, S. S., & Ramesh, V. (2016). Demand side management scheme in smart grid with cloud computing approach using stochastic dynamic programming. *Perspectives on Science, 8*, 169–171.
30. Parolini, L., Tolia, N., Sinopoli, B., et al. (2012, 13–15 April). A cyber-physical systems approach to energy management in data centers. In *Proceedings of the 1st ACM/IEEE international conference on cyber-physical systems, Stockholm* (pp. 168–177). New York: ACM.
31. Zhao, P., Simoes, M. G., & Suryanarayanan, S. (2010 8–10 Nov). A conceptual scheme for cyber-physical systems based energy management in building structures. In *2010 9th IEEE/IAS international conference on industry applications (INDUSCON), Sao Paulo, Brazil* (pp. 1–6). New York: IEEE.
32. Zhao, P., Suryanarayanan, S., & Simoes, M. G. (2013). An energy management system for building structures using a multi-agent decision-making control methodology. *IEEE Transactions on Industry Applications, 49*, 322–330.
33. Paul, A., Jiang, Y. C., Wang, J. F., et al. (2012). Parallel reconfigurable computing-based mapping algorithm for motion estimation in advanced video coding. *ACM Transactions on Embedded Computing Systems, 11*(S2), 40.
34. Lee, J., Bagheri, B., & Kao, H. A. (2015). A cyber-physical systems for industry 4.0-based manufacturing systems. *Manufacturing Letters, 3*, 18–23.
35. Paul, A., & Rho, S. (2016). Probabilistic model for M2M in IoT networking and communication. *Telecommunication System, 62*, 59–66.
36. Paul, A., Victoire, T. A. A., & Jeyakumar, A. E. (2003, 27–30 Dec). Particle swarm approach for retiming in VLSI. In *Proceedings of the 2003 IEEE 46th Midwest symposium on circuits and systems, Cairo, Egypt*. New York: IEEE.
37. Bhushan, D., & Agrawal, R. (2020). Security challenges for designing wearable and IoT solutions. In *A handbook of Internet of Things in biomedical and cyber physical system* (pp. 109–138). Cham: Springer.

Chapter 11
Solar-Powered Portable Charger Using IoT-Based Technique

Monika Sharma and Abhijeet Chauhan

11.1 Introduction

As per continously degrading of natural resources like fossil fuels in order to complete the demand of the public and different business and private sectors for the use of electricity, the security of natural resources in order to produce energy is to be maintained to maintain a balance and sustain the energies. The power play can be brought into the implementation by using a resource that has limitless power; the renewable energy are capable of playing a major role to sustain power. The clean and abundant amount of solar energy will help to combine with the electricity sector to offload the pressure. In India the solar energy plant and operations are co-operated by the *Ministry of New and Renewable Energy* also initialized as MNRE. A huge amount of solar-powered energy is anticipated by 2022, an amount of 100 gigawatt. The solar energy is fuel independent, needs no maintenance, and is expected to be long-lasting.

The predicted scheme for the provision of electricity to the backward areas and villages from renewable and standard source on independent basis is the *Decentralized Distribution Generation (DDG) scheme*. For fulfilling the schemes, a low, DC distribution voltage network will be used for obtaining the low energy consumption to remote areas where the solar home system process can be interconnected to each other for covering various households and will lead up to an attractive option. This chapter will drive through its way from the architecture to the development of the SHS efficient solar-powered battery system programmed by a microcontroller, Arduino UNO.

M. Sharma (✉) · A. Chauhan
AIIT, Amity University, Noida, UP, India
e-mail: msharma5@amity.edu

M. Moh et al. (eds.), *Smart IoT for Research and Industry*, EAI/Springer Innovations in Communication and Computing,
https://doi.org/10.1007/978-3-030-71485-7_11

11.2 Standard Methodology

The solar panels of the Photovoltaic System's major independent component are the power detection and condition service with the batteries. Specific convertors are used in order to change the energy that the device stores from DC to AC. The convertor makes the whole module complex and leads to less efficiency of the power system to be used effectively. At times the power in the battery while converting is lost in high percentages because the power has to be converted back from AC to DC because now mainly major electronic devices like mobile phones, laptops, and other devices use DC power for its working [1].This module is efficient in places where the population is sparse, and low-voltage devices can help for the generation of the electricity produced by the PV module in order to supply solar energy, and this amount of energy can be used in DC devices like LED lamps, TVs, mobile charging, and DC fans. To reduce the complexity and for a cheaper and reliable model, the power system should be represented for an updated and efficient network of power like represented in Fig. 11.1.

This system is effective because the power system complexity is reduced as DC current eliminates the usage of convertor and as well as convertor cost, and thus reliability is increased. Solar panels use the batteries as its energy bank of emergency that it stores for applications to be used in local and industrial places. The amount of sunlight at a constant rate is required for storing the energy in these batteries, but the batteries are sensitive as they need special care for their long-lasting serving; these batteries cannot be supplied with extra current and cannot bear the energy if overcharged. All the factors will lead the battery to die at a high pace. Another instruction for the batteries is that when a battery is discharged deeply, then the chemical reaction to kick back the energy is required, and it is obtained through passing large current. The batteries should not be left for longer periods of time for processing the charging; the overcharge will lead to oxidation of the battery and will further react with hydrogen, and the plates that are attached to cathode and anode will fail to resist and lead to failure in battery. So, a small DC-supported model is aimed that is cheap and reliable.

The figure down below presents the architecture through a diagram of the system to be proposed. In whole, *the independent PV system powered by solar energy*

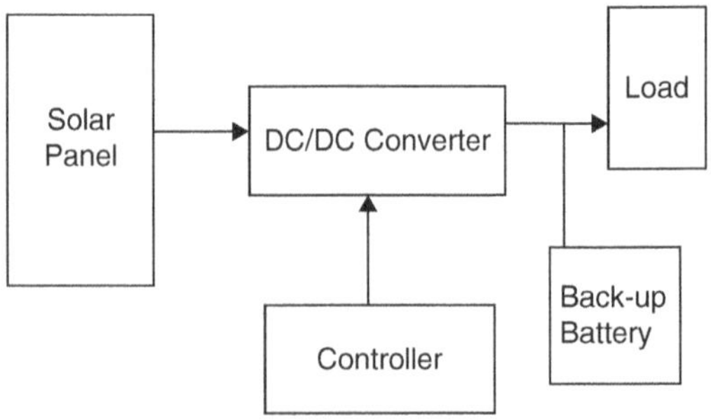

Fig. 11.1 Standard methodology

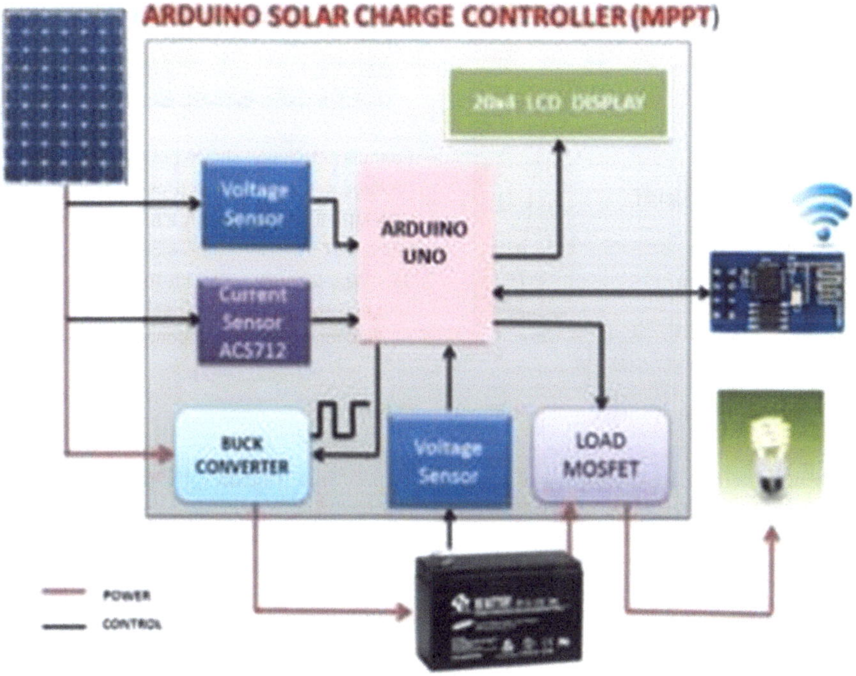

Fig. 11.2 Arduino solar charge controller (MMPT)

architecture includes an Arduino microcontroller that interfaces MPPT algorithm, battery, solar panels, Wi-Fi module, and an output source to supply the energy stored within itself (Fig. 11.2).

11.3 Efficient Working of the Model

The Architecture of The Desired Model Explained.

11.3.1 Photovoltaic System Cell

A PVC module can be explained as an arrangement of 5 × 10 solar cells connected together that is joined together. The PVC system of the solar array is generally comprised of the solar electricity of the photovoltaic cells and is utilized in regular or domestic application [5]. The solar panel consists of a number of silicone-based PV cells combined in series and in parallel depending on the required voltage and the

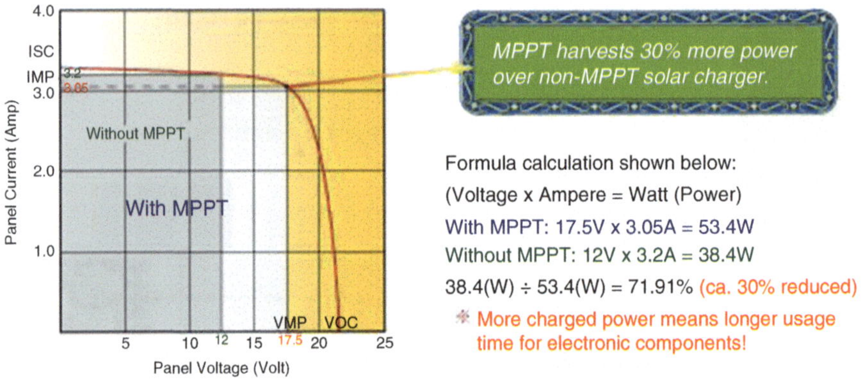

Fig. 11.3 Current and voltage graph (with\without MPPT)

current. Few precise examples of PVC systems of an array of solar modules are as follows:

- Inverter powered by solar power.
- A lithium-ion battery.
- An energy tracker with interconnected wiring.

A. **Voltage of the panel: 12 volts.**
B. **Efficiency of each solar cell: 17%.**
C. **Efficiency of the module: 14.6%.**
D. **Short circuit current (in amp): 3.32 A.**

The characteristics of the solar cells under the sunlight, from A to D or I to V points, define the knee point where the optimal power is available in order to analyses the intersection of power line with the dropping voltage line. As seen in the figure below (Fig. 11.3):

11.3.2 Batteries for Energy Storage

When the obtained energy is to be stored, there are applications that require this stored energy, and that includes the energy and power application, both treated differently as energy sources; when compared the discharging of power application is faster than the energy applications. In energy application it varies from 10 min to hours, whereas in power application the rate of discharging is very high, and it varies from seconds to minutes and thus leads to shortage of energy. Another difference is that in power applications, the media to store the energy obtained through solar power are capacitors, battery, and hydrogen. Lead acid batteries have an advantage of superior power ratio over rechargeable batteries used for the model.

11.3.3 Controller of the Charge

Any item after exhausting the resources needs a recharge, and this model is not complete without talking about its recharge. The device sensor attached is used for restricting the overcharge of the batteries, and the controller helps to convert the current from a solar panel to the required DC format for functioning. The controller comes with an auto-cut feature that restricts the battery from getting overcharged. Other features include when the charge is being converted to DC format, the intensity of the light is backed up by the controller and does not completely shuts down which regains and restores back again. The energy stored as a charge in the battery uses *maximum power point tracking (MPPT)* that helps to gain control over the voltage that prevails maximum light conditions.

The algorithm and controller for the circuit uses the MPPT technique, which is costly but reliable and affective.

11.4 A Short Description of MPPT Algorithm

The algorithm of maximum power point tracker (MPPT) is explained with the help of block diagram. *Pulse width charge initialized as PWD* is used by the controller in order to control the current from pulses that are of high frequency; the state of the battery is a matter of concern while applying the technique. PWD plays a major role to restrict the battery from overcharging. While the usage of the batteries when they tend to discharge the PWM sensor known as *pulse width modulation* sends a signal [4]. After the signal is received, the pulses with full current passing remains to be continuous, and simultaneously the batteries perform bulk charging side by side. The next stage occurs when the batteries are at a stage near to full; the *stage of charge (SOC)* of the diagram of charging includes the step of *absorption*. The controller has the power to control the flow of battery in the module and thus do so. The capacity of the batteries when full is called *floating charging stage* as per the block diagram. The different battery charger phases include different states (Fig. 11.4):

- **On state:** This stage is the initial stage in which the charge of the solar panel is valued and observed to be between minimum or low solar power values.

$$\textbf{Minimum solar power} < \textbf{solar power} < \textbf{low solar power}$$

The bulk charging state kicks in when the input watt is very low; the PWM is set to 99.9% in order to prevent the bulk charging to turn off completely. The PWM helps in recharging and bringing back the energy.

- **Bulk state:** This is a state of charge in which solar power comparatively lower down than the current solar power. This states include the PPT algorithm which

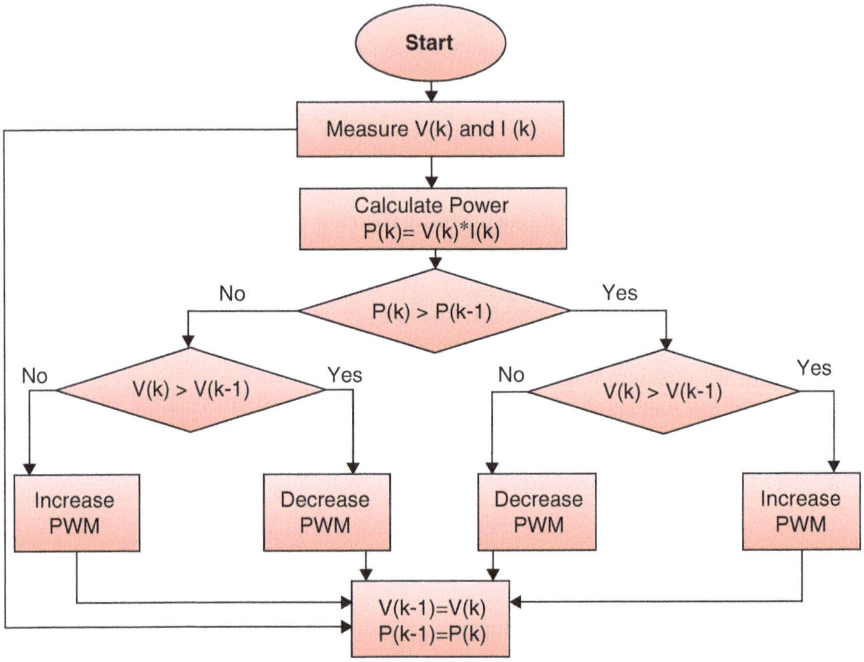

Fig. 11.4 Flowchart for MPPT algorithm

runs the maximum amount of current in the circuit that helps in bulking the battery by the charge.

- **Float state:** The voltage in this stages rises to the level where maximum voltage is achieved, and PWM value is adjusted in the state to maintain the module to maximum voltage. The voltage is kept maximum for a period of time till the battery charges up to 100% because if further the voltage remains maximum then it will lead the battery to lose its power. This state is called the float state.
- **Off state:** when the solar panel is unable to generate power, then the model automatically goes to off state. Which implies:

$$\textbf{Solar power} < \textbf{minimum solar power}$$

The power from the solar panel used in the module is reflected to the display; all the *metal–oxide–semiconductor field-effect transistors* (MOSFETs) are set to off to avoid full drainage of battery. *The final module of Arduino circuit* was used with the MPPT technique to overcome the problem that when at night the solar power is producing no energy but the *load state* keeps on obtaining energy from the panel even if the panel has no energy to give.

11.4.1 Convertor (DC to DC)

To boost or increase and buck or decrease of current when converted DC to DC requires the voltage panel to be at a targeted battery level. The circuit that uses different parameters for the decrease of current include *inductor, capacitors, and MOSFETs*. In particular in this module the MOSFET IRFZ44N is used for increasing or decreasing a current, thus acting like a switch [6].

11.4.2 Arduino UNO

The microcontroller is easy to program as it is user-friendly, and the architecture of the Arduino that is used has 12 pins that are mainly used for power to the circuit and other connections of components to the circuit that will work accordingly as programmed. The input/output, analog input/output, and control pins are used [2] (Fig. 11.5).

Number of total pins used: 14 pins
- [Analog] Number of pins used as PWM inputs: 6 pins.

[Analog] Number of pins used as PWM output: 6 pins
USB connector to computer: 11th pin
Operating voltage of 5 V: 12th pin.
A reset button.

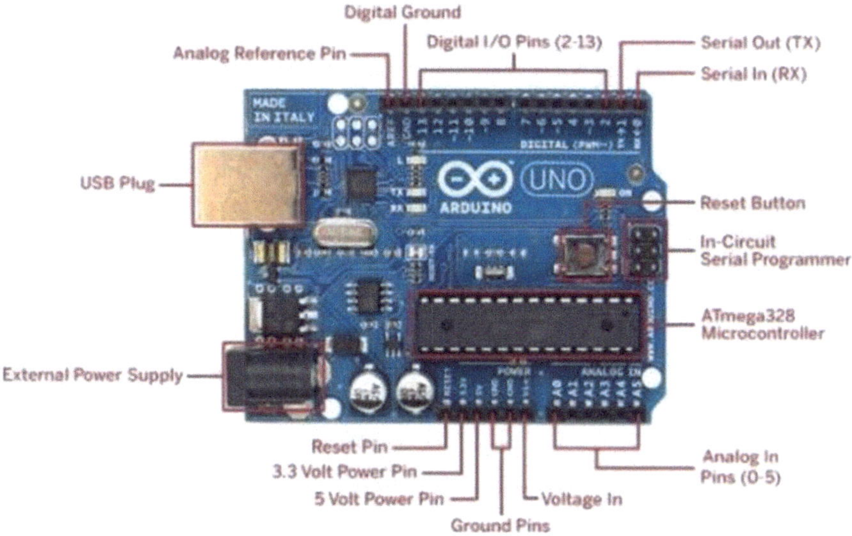

Fig. 11.5 Arduino UNO

11.5 Arduino and Wi-Fi Module Interface

The model for connection of the Wi-Fi module is the ESP8266 which is used in accomplishing and monitoring the task; it is a self-controlled system on chip that supports the integration of TCP/IP module which helps any microcontroller to Wi-Fi network. The functions of the Wi-Fi module helps to offload the application or host it from some other application. The ESP8266 Wi-Fi module is initially programmed with automatic command set and microcodes that help in the execution for the connection and different uses of Wi-Fi. The programming when completed is executed by serial adapter and the other end connected to USB that helps to burn the code in the microcontroller [3]. Like in the circuit figure below, the connections are easy and clear. The microcodes to be programmed in the microcontroller without any USB interface use the ESP8266 module that transmit program with wireless communication with the module; further the flashers are used to flash the microcodes into the ESP module. Through automated commands, the data sharing is done through the Internet from the module to the microcontroller. After that for the data logging, the photovoltaic system attached to the battery is used for various purposes (Fig. 11.6).

Even for longer distant commands, the module is effective, and the information can be sent to nearby locations of the solar tree system through mobile phone transmitter and received by the Wi-Fi module.

All these features lead to a better and efficient designing of the IoT module that is reliable and easy to connect.

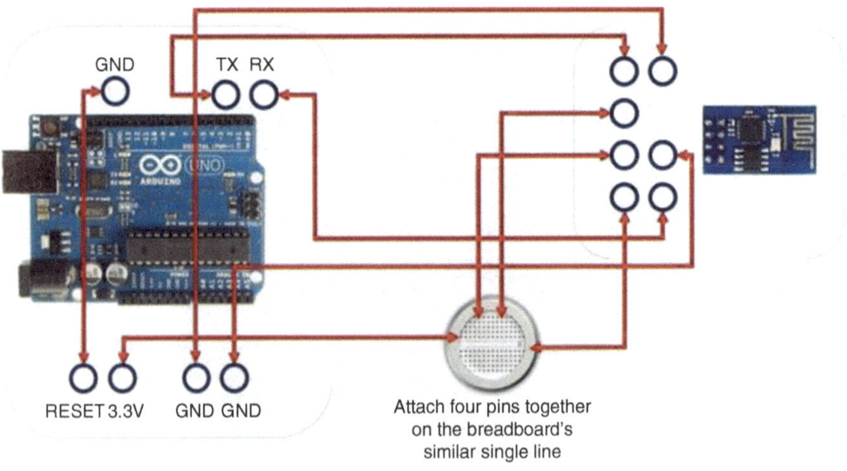

Fig. 11.6 Connection between Arduino and WiFi module

11.6 Tips and Tricks to Keep the Power Bank Safe

To get the best use from any PB, there are a few tips that will help ensure the work rate is gained from the power bank for a long time [7].

- **Maintain optimal temperature:** It is not always possible that the lithium rechargeable battery technology used in PB performs best at normal temperature. It does not like to get cold or too warm either. Both is not good for lithium rechargeable.
- **Charge before using it:** Manufacturers always suggest giving a new PB a full charge before use.
- **Battery powered up:** This may sound obvious, but the power bank is no use if it is not charged up. It helps to get used to charging it up as soon as possible after it has been used. In this way it will always be ready for use.
- **Charge the PB when not in use:** Rechargeable batteries do not like being discharged for a long time. As these batteries always lose a little battery with time, it is best to recharge the power banks from time to time when they have not been used for some time.
- **Use it in a calculated way:** A PB should be only used for charging the type of device for which it is compatible with. Generally, devices have different voltages, and if they have a USB port this would not be the problem.
- **Put it away from being wet:** These things are electronic devices, and that is why they do not like water. One main advice when using power banks is to keep them always dry. With the amount of power within the power bank, moisture can cause significant damage.
- **Keep away from the contact of metallic things:** The amount of power in PB is the potential of short circuit, and one of the important things of using bank is to make sure they are never shorted.
- **Do not lose it by dropping:** PB includes circuits and the battery. Like every other electronic object, they need to be handled with some protection. Dropping them can damage the circuit board, the case, or even the rechargeable battery.

PB rechargeable batteries can be easily used. With a few rational safety measures, they really provide extra charge when you are on the go. Take care when handling them.

11.7 Conclusion

The chapter defines the architecture and update to the design of the microcontroller in an efficient and powerful way that helps to analyze, keep a check, and observe other changes in the module that is powered by solar energy and further saved in batteries. The charging of batteries is controlled through the microcontroller by the MPPT algorithm. The charge control comes with extra features that include the

Wi-Fi transmission and a display to tackle and observe the output of each step taken. The automated voltage is tackled by the automatic module transmission which restricts the battery to overcharge and undercharge and helps to last long, thus making the whole system reliable and efficient with the remote surveillance of the independent battery-operated microcontrollers.

References

1. EL-Mohan, M. S. (2012). Fuzzy controller design using FPGA for photovoltaic maximum power point tracking. *International Journal of Advanced Research in Artificial Intelligence, 1*(3).
2. Zhou Yangzhou Jinxing. (2010). *Application of Fuzzy Logic control approach in a microcontroller based sun tracking system*. WASE international conference on Information Engineering.
3. Arduino connection with Wi-Fi circuit. http://www.teomaragakis.com/hardware/electronics/how-to-connect-an-esp8266-to-an-arduino-uno/
4. MPPT VS PWM. https://www.youtube.com/watch?v=PB6zojol9o0
5. Maunder, J., Naraseeyappa, S., & Antalkali, S. (2017). Analysis of PV techniques to avoid loss of energy and resources: Application of saving the expensive PVC module and attached solar panel. *J Big Data, 4*, 20. https://doi.org/10.11106/s405437-0157-00677-74.
6. AC to AC or DC to DC charge convertor. https://wallbox.com/en_catalog/faqs-difference-ac-dc#:~:text=The%20difference%20between%20AC%20charging,onboard%20charger%20to%20convert%20it.
7. https://www.electronics-notes.com/articles/equipment-items-gadgets/powerbank/how-to-use-powerbank-tips-dos-donts.php

Chapter 12
CBC (Cipher Block Chaining)-Based Authenticated Encryption for Securing Sensor Data in Smart Home

S. Rajashree and R. Sukumar

12.1 Introduction

As technology improvement is happening rapidly, it is making the human daily life easier with the use of newly developed efficient smart automated systems. Due to the fast and improved Internet technology and advancement in microcontrollers, microprocessors, and smart embedded systems [1], people are more motivated to use the Internet for controlling and adapt to different types of advanced automated devices. Human-machine interaction (HMI) is becoming more realistic in everyday life due to this technology upgrade. Today, this upgrade in HMI research has moved towards using things connected to the Internet, which was previously used for communication of data between computer and humans; now it is used for things, and here things may be physical devices that are connected to sensors and actuators or virtual things, i.e., IoT (Internet of Things). The aim of this chapter is to connect any things physical or virtual which can communicate between themselves by use of the Internet that can be accessible from anywhere [2]. The Internet of Things (IoT) creates an innovative improvement in the technology world with a new era where intelligence computing can be achieved using existing platform.

S. Rajashree (✉)
Department of Computer Science, Jain University, Bangalore, India

R. Sukumar
EC Department, Jain University, Bangalore, India
e-mail: r.sukumar@jainuniversity.ac.in

M. Moh et al. (eds.), *Smart IoT for Research and Industry*, EAI/Springer Innovations in Communication and Computing,
https://doi.org/10.1007/978-3-030-71485-7_12

12.2 Related Works

IoT in simple terms is defined as the connection between many kinds of end devices like smartphones, personal computers, and tablets through means of the Internet, which brings in the very new method of communication between things and people and to data storage in a very secure manner. Here, the main goal is to achieve control over any kind of electric and/or electronic objects or devices surrounding us in an easier, innovative, smooth, and secure way. One more objective of this technology improvement is to achieve prominence performance of electric devices by confirming cost-effective output with more protection and safety and also to provide entertainment. IoT technology is used to come in with an idea and also to achieve great growth for smart automated homes that aim to achieve higher living standards of humans.

Nowadays the concept of a smart home [4, 14] is a becoming interest among most of consumers as it is leading to easier maintenance of end devices. Here in this chapter, a discussion about a system for home automation and metering is proposed. A website can be designed, which can help home appliance and other electronic end devices to be controlled through a website from a remote location by using the Internet. Home automation here means the use of computer and information technology combined with physical and virtual things to control most commonly used smart appliances and features (such as movement of windows or smart bulb). Secure systems may be a simple remote control of smart bulb or a complex microprocessor/ microcontroller (such as Arduino board [6])-based networks with higher degree of intelligence and automation as in case of driverless smart cars. Automation for each of smart application is used for a number of reasons like ease of operating, more enhanced security, and achieving more efficiency for energy utilization [7]. This chapter discusses that home automation can play a vital role for better energy management using a renewable energy sources.

IoT is a system of connected interrelated intelligent computing devices and virtual and digital machines [3] that connect objects, animals, or human beings that can be provided with unique identifiers which can generate enormous amount of data and can transfer data over an Internet without interference of human or computer systems.

The main objective of the IoT is to make things to be connected and communicating anytime, anyplace, anything, and anyone who aims at achieving communication using any path/network from any service provider. The Internet of Things helps humans to live and work smarter with affordable cost and in fewer amounts of time, as well as gain complete control over their lives. In addition to offering smart automated and efficient devices for homes, IoT is essential for commercial benefits. IoT has various types which help in businesses with real-time systems and to look into deeper view of systems and provide analysis so that one can aim at designing a more efficient system to supply chain and logistics operations.

When IoT involves generating and communicating enormous amount of data, definitely security should be given to the data where data is generated during communication and also where data is stored [5]. Security for data can be provided by encrypting the data. Data can be authenticated by using message authentication codes. So by combining encryption and authentication, more protection can be given to data against adversaries.

In symmetric system, there will be one key used between the sender and the receiver. The sender and the receiver share this secret key with each other by some secure means. In contrast, in asymmetric key cryptosystem, the sender generates a pair of keys (public, private) which have some relation between them. The private key will be retained by the generator who might be the sender of the information, and the public key will be shared to those who needs to decrypt the value that is sent by the sender to get the plaintext or can use public key to send some information to the sender who can decrypt using the private key.

12.3 Introduction to AES

NIST hosted a symmetric key encryption competition. The Rijndael cipher ultimately won from other finalists such as RC6, Serpent, MARS, and Twofish. It became the advanced encryption standard (AES). Its creators are Joan Daemen and Vincent Rijmen. The reason for the replacement of DES was that DES could be brute-forced easily. This is not safe as anyone with time and computation power could render DES useless. Since DES was made by IBM for the NSA, this security flaw could not be overlooked, and a replacement for DES was sought after. Rijndael was chosen as the AES due to being easy to understand and efficient and having low memory usage. One secure method in symmetric key system is the advanced encryption standard method (AES) [9]. AES uses 128 bit key for encryption. This algorithm has particular structure to encrypt and reverse structure decrypt sensitive data and is implemented in both hardware and software. It is extremely difficult for adversaries to get information about plaintext or information or key used when the data or plaintext was encrypted by AES algorithm. The cryptanalysis of AES is very difficult as it has avalanche effect with cipher text. AES has three variations with key sizes which are AES 128, 192, and 256 bit, and the block size of AES design is 128 bit block.

AES design principle is based on substitution–permutation network, which is an efficient polynomial algorithm implemented in both software and hardware. Unlike DES, AES does not use a Feistel network that gives chances to adversaries to perform and has linear and differential cryptanalysis on S-boxes and round functions. AES is an upgraded version of Rijndael cipher, with a fixed block size of 128 bits and a key size of 128, 192, or 256 bits [11] (Fig. 12.1).

AES is dependent on SP network or substitution–permutation network (SPN) concept. SP network is cryptographic primitives formed from mathematical operations used in block cipher AES algorithms where s is for substitution function and p is the permutation network. Permutation operation is required to remove any regular patterns which may appear in the cipher text (i.e., diffusion), while substitution operation is needed to achieve cipher text ambiguity goal (i.e., confusion).

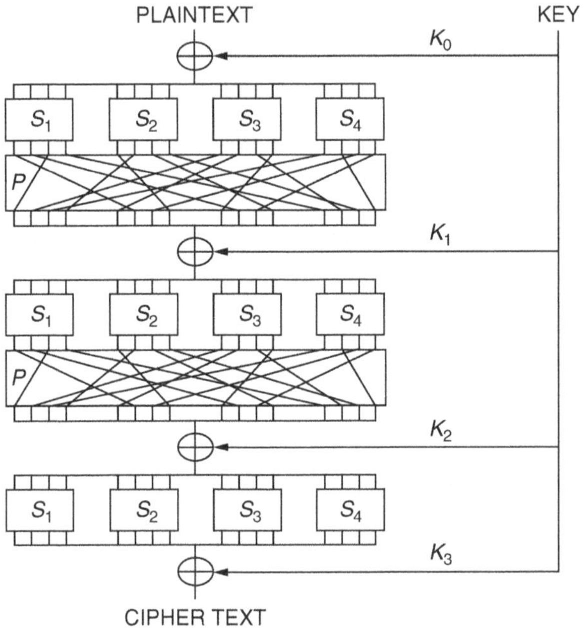

Fig. 12.1 SP network

Diffusion is the cryptographic term that defines the relationship between plaintext and cipher text where one bit value change in plaintext leads to minimum 2 bit changes in outputted cipher text, and then all these changes are distributed by the P-box among several S-boxes, so that the outputs of all S-boxes are changed at several bits. Doing several rounds of SP operations ensures that bit changes several times back and forth; therefore, at the end, the resulting cipher text has complete variations when compared to input bit value.

Confusion is a cryptographic term that defines the mapping between the key and cipher text making this mapping complex very important for achieving secret system. Changing one bit of the key value should reflect major changes in the cipher text in a very complex manner.

Even if an attacker somehow obtains plaintext by some means corresponding to a part of cipher text—the adversary getting the whole plaintext with the existing knowledge about plaintext should not learn more about the remaining cipher text, that is the system should provide security with respect to COA, CCA, KPA, or CPA models. This is achieved by using confusion and diffusion concept.

12.3.1 Constraints and Assumptions

Once the receiver receives encrypted data, the seed receiver begins the decryption process which is reverse of encryption process as all functions used in AES are invertible functions. The structure of decryption process is slower than the

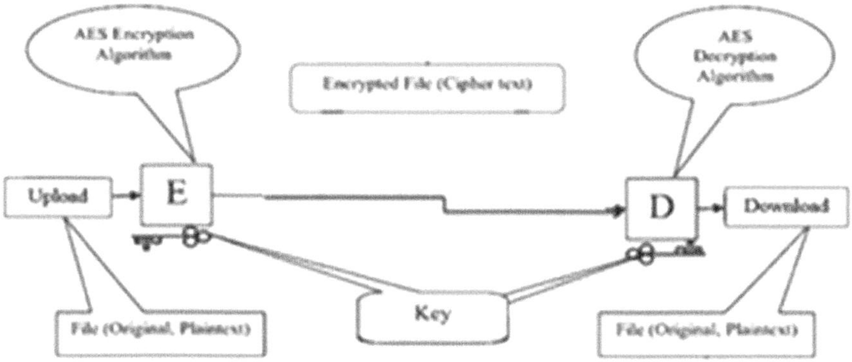

Fig. 12.2 AES encryption and decryption

encryption process because of this unbalanced structure in encryption, and decryption process have different polynomial time complexities (Fig. 12.2).

AES cipher with secure of 128 bit is more appropriate for small data generating applications like IoT sensor data. Therefore, these applications with small amount data may not need a larger algorithm with a larger order of mathematical and structure foundations. Because of this, AES proves to be secure and appropriate algorithm for securing IoT data.

12.3.2 Dependencies

AES encryption process depends on the fixed S-box having a mapping of 16×16 values in hexadecimal notation and in decryption process also has an inverse substitution matrix that can give the plaintext upon mapping in reverse.

The AES has a defined number of rounds that depends on key. Each of these rounds uses a different 128 bit round key, which is calculated from the original AES key. The algebraic structure is simple, and each block is encrypted in the same way so if the hacker is able to decrypt any of the blocks, it becomes easy to decrypt the rest of the blocks (Fig. 12.3).

12.3.3 Evaluation Parameters

- The CBC-AES cipher that is implemented [10] in hardware using parallel implementation will execute around 23 times faster than the AES cipher implemented as software.

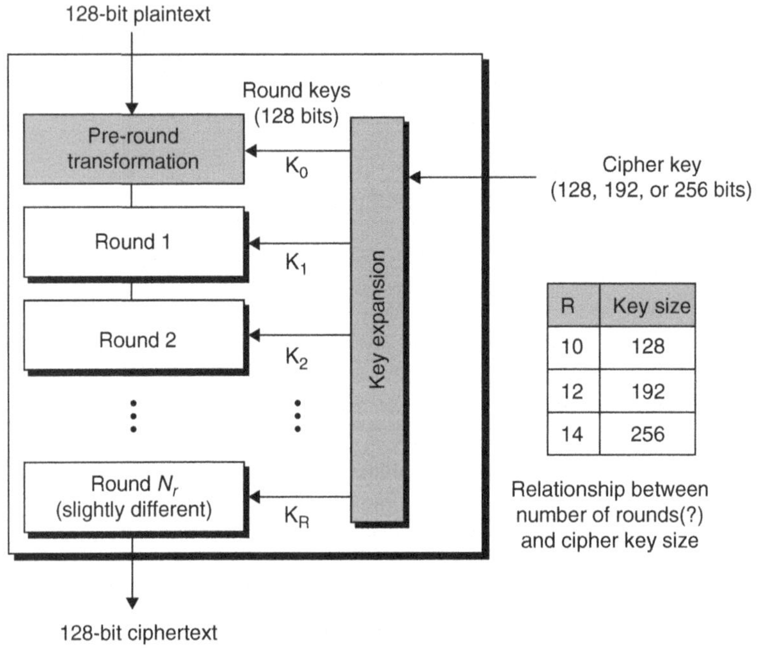

Fig. 12.3 AES rounds and key size

- The CBC-AES inverse cipher if implemented in hardware utilizing parallel implementation concept will execute around 29 times faster than the CBC-AES inverse cipher implemented in software [13].
- Similarly even the key generation and expansion function when implemented in hardware runs faster when compared to software implementation.

A lot of data that is generated by IoT end devices are sensitive and personal data that needs to be protected from intruders. Encryption algorithms play very vital roles to shield data from unauthorized access. Numerous efficient polynomial algorithms exist for these problems. Advanced encryption standard (AES) algorithmic program is one among the most efficient polynomial algorithms, and it is widely supported and adopted on hardware and packages. This algorithm enables to encrypt data in three completely different key sizes like 128, 192, and 256 bits with 128 bits block cipher [12].

12.3.4 Working of AES

For encrypting sensor data AES 128 is used. AES 128 takes 128 bit key and operates on 128 bit block data. 128 bit key is given to key scheduling algorithm which gives sub-key for each round. For generating sub-key, it uses a substitution table along with some constant value r-con. AES is mainly used in the protection of data and is even used by the US government to protect sensitive data. AES is more mathematically elegant and is of simplistic design than the previous DES. It provides various key sizes, even the smallest of which is not practical to brute-force. Some drawbacks of the algorithm is that the inverse of the algorithm is a different code and the circuitry used for encryption cannot be reused. AES uses the following functions for its implementation: ByteSub function, ShiftRow function, MixColumn function, and Round Key addition. Each function is explained below ByteSub uses the S-box (lookup table) to substitute the bytes in the block from those in the lookup table. It is a nonlinear transformation. Nonlinear transformations can decrease or increase correlations between variables. For example 14 gets replaced AES S-box. Column: least significant nibble. Row: most significant nibble. ShiftRow cyclically rotates each row by an increasing number of bytes. The zeroth row has no shift, the first row has one shift, the second row has two shifts, etc. 128 bit data and 128 bit sub-key will be represented as state. The state is a 4×4 matrix, and each is of 1 byte value. So data in one block is 4*4*8 = 128 bits.

The process of AES mainly involves four steps:

- AddRound Key.
- Substitution and inverse substitution.
- ShiftRows and inverse ShiftRows.
- MixColumn and inverse MixColumn.

AddRound key is the step which is performed with all the ten rounds of AES before data given to substitution step AddRound key is performed. This involves performing XOR on each bit of plaintext with each bit of sub-key generated by key scheduling algorithm. The output of AddRound key which is also a state 128 bit will be given as input to substitution step.

The substitution step consists of substitution table (S-box sown in Fig. 12.4) which has an entry of 16×16. Each byte value of state is divided into two parts starting with 4 bits; 4 bits constitute the row value, and the next 4 bits form the column value. A value of 8 bit is outputted using row and column value as input to the S-box. The substitution step will produce an output of 128 bit state. This is given as input to the next step which is ShiftRow function.

The ShiftRow function shifts the second row of state by 1 byte, the third row by 2 bytes, and the fourth row by 3 bytes. This is not a binary bit shift but it is a whole byte shift. The output of AES ShiftRow is also a state of 128 bits, and this becomes input to the last step of AES round that is MixColumn. The MixColumn step is performed in almost all rounds except the last round in AES 128; the tenth round is missing the MixColumn step. The final output after MixColumn is given to AddRound key function for second round after ten rounds of 128 bit state is generated which has an avalanche effect on 128 bit input data. In the receiver side, all steps are done in the reverse process. AES all function are invertible function that is:

$$f^{-1}\left(f\left(x\right)\right) = x.$$

	0	1	2	3	4	5	6	7	8	9	A	B	C	D	E	F
0	63	7C	77	7B	F2	6B	6F	C5	30	01	67	2B	FE	D7	AB	76
1	CA	82	C9	7D	FA	59	47	F0	AD	D4	A2	AF	9C	A4	72	C0
2	B7	FD	93	26	36	3F	F7	CC	34	A5	E5	F1	71	D8	31	15
3	04	C7	23	C3	18	96	05	9A	07	12	80	E2	EB	27	B2	75
4	09	83	2C	1A	1B	6E	5A	A0	52	3B	D6	B3	29	E3	2F	84
5	53	D1	00	ED	20	FC	B1	5B	6A	CB	BE	39	4A	4C	58	CF
6	D0	EF	AA	FB	43	4D	33	85	45	F9	02	7F	50	3C	9F	A8
7	51	A3	40	8F	92	9D	38	F5	BC	B6	DA	21	10	FF	F3	D2
8	CD	0C	13	EC	5F	97	44	17	C4	A7	7E	3D	64	5D	19	73
9	60	81	4F	DC	22	2A	90	88	46	EE	B8	14	DE	5E	0B	DB
A	E0	32	3A	0A	49	06	24	5C	C2	D3	AC	62	91	95	E4	79
B	E7	C8	37	6D	8D	D5	4E	A9	6C	56	F4	EA	65	7A	AE	08
C	BA	78	25	2E	1C	A6	B4	C6	E8	DD	74	1F	4B	BD	8B	8A
D	70	3E	B5	66	48	03	F6	0E	61	35	57	B9	86	C1	1D	9E
E	E1	F8	98	11	69	D9	8E	94	9B	1E	87	E9	CE	55	28	DF
F	8C	A1	89	0D	BF	E6	42	68	41	99	2D	0F	B0	54	BB	16

Fig. 12.4 AES S-box

- The key space of AES = $\{0,1\}^{128}$ so the probability with getting a key using brute-force attack is very negligible = 1/2128. For processing each block of data, cipher block chaining (CBC) [8] mode of operation is very useful in achieving higher security and authentication.

12.4 Mode of Operation: Cipher Block Chaining

CBC mode of encryption provides cryptographic security wherein every encryption of the same plaintext results in different ciphertexts. This is achieved by what is known as the initialization vector (IV). The initialization vector (IV) is required to be of the same length as the block that is being encrypted. The various steps involved in the encryption process are as follows:

1. The plaintext block (P1) is XORed with the IV.
2. The output from step 1 is encrypted with the key (K).
3. The results of encryption on each block (C1, C2, …, CN-1), as specified in steps 1 and 2, will be used in XOR operation for the next plaintext block (Pn).

This is how different ciphertexts are obtained with every encryption of the same plaintext. It is to be noted that the same key (K) is used in each of the encryption blocks (Fig. 12.5).

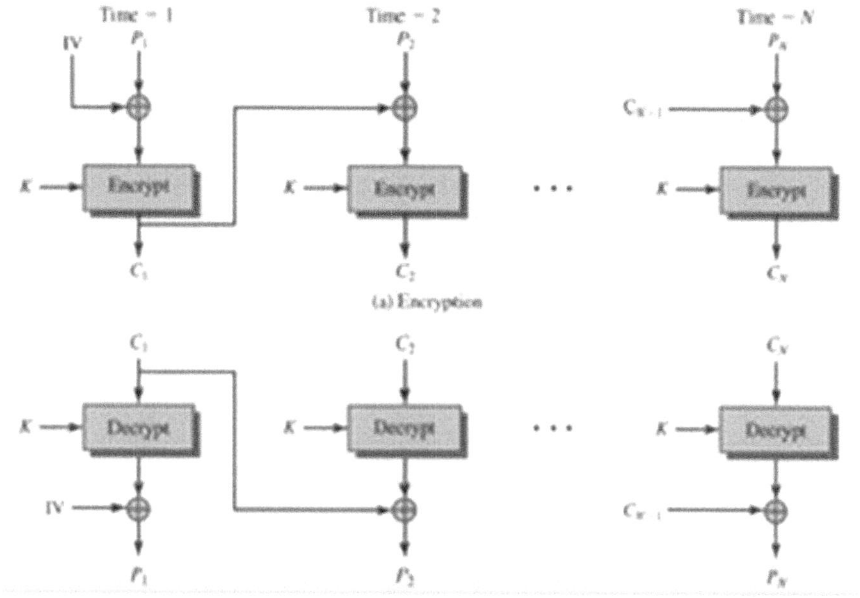

Fig. 12.5 CBC mode of AES encryption

Here, the previous ciphertext block is added to the current ciphertext block, and the result is encrypted using the key. Decryption can be thought of as a reverse process wherein the current ciphertext is decrypted using the key, and then the previous ciphertext block is added to the result. It is worth noting that this mode (CBC) forms the basis for data origin authentication or data integrity.

AES is the most used symmetric algorithm in the world and therefore is used wherever security is needed. Below are some of the applications of AES:

A. Data protection: AES was originally made to keep classified data in the US government secure. Although it started its roots in the government, due to being open source, AES is one of the most used algorithms for data storage commercially and privately. Dropbox uses AES 256 to encrypt files and store them while using SSL when transferring them.
B. Data transfer: AES is one of the algorithms which is used in https. It is also used in the SSL layer.
C. Image encryption: Variations of AES such as those that we have discussed earlier use chaos to increase the diffusion of the pixels therefore reducing any vulnerabilities that can be effected by confusion-diffusion analysis and/or histogram analysis. We have discussed a variation that uses six chaos maps to protect and transfer satellite images safely.
D. IOT AES is used to encrypt the channels between wireless devices. The most common example is the WPA2 protocol which combines with AES to create the WPA2/AES protocol which is much safer than previous versions.

AES has several variant algorithms with respect to the cardinality of key and round count for encryption which are detailed as below.

(a) AES 128 is an AES variant that uses 128 bit blocks and 10 rounds. Relatively least secure and has the shortest key of 128 bits. Adversary needs 2 different key combination trials. High 128 speed computation with less computational power. Preferred when encryption needs to be at high speed due to its low complexity.
(b) AES 192 is an AES variant that uses 192 bit blocks (block size not adapted by the AES standard) and 12 rounds. More secure and longer key than AES 128 of 192 bits. Adversary needs 2192 different key combination trials. Slightly slower computation with high security. Most preferred AES variant for data encryption.
(c) AES 256 is a AES variant that uses 256 bit blocks (block size not adapted by the AES standard) and 14 rounds. Most secure and longest key of 256 bits. Adversary needs 2 different key 256 combination trials. It uses up to 40% computational power and resources with high complexity and more making it slow but provides high security. Preferred in high-security environments with extremely sensitive information.

12.5 Attacks on AES

12.5.1 Side-Channel Attack

This attack aims to collect information on the execution of the algorithm and not on breaking the cipher directly. Since lookup tables are used in AES to increase the speed, cache misses/hits are common as these tables might not fit into the cache fully. This varies the time it takes for encryption using different keys. The timing info, along with info such as cache contents and power consumption, is used to guess the key. To protect against this attack, two types of countermeasures are usually investigated:

- Preventing information from leaking.
- Introducing noise to the leaked info to prevent any correlation between leaked information.

12.5.2 Differential Fault Analysis

Differential fault analysis is an attack used to break block ciphers. It compares the differential information between correct and incorrect cipher texts. To get faulty cipher texts, we provide an external voltage variation, glitch, etc. DFA can be done on either the state or the key schedule.

Improved DFA: An improved DFA will be discussed here. This method makes DFA attacks more efficient in the AES key schedule. The advantages of the improved method are as follows:

- The improved attack uses minimal faults as they fault the key scheduling therefore further faulting the state.
- It uses a one-byte fault model.
- Information on what the plain text was is not required. To implement this attack, we introduce a fault at the first column of the eight round of AES key schedule by applying a voltage hit to it. We do not know what the corrupted value is.

This fault carries over to the Rotation and ByteSub function, and the transformations are done to compute the new round key. We use differential equations at each round to reduce the number of unknown key parts. The fault propagation during the last three rounds is shown. To prevent DFA attacks generally, the last rounds are recomputed and compared to see if any fault was there.

12.6 Wireless Sensor Networks

A sensor node, which is also called as mote, is an end node in a sensor network. Sensor node can do few processing, gathering sensitive data, and communicating with other connected nodes in the sensor network. A mote is a node, whereas a node is not always a mote.

The main components of a sensor node are as follows:

- Microcontroller.
- Transceiver.
- External memory.
- Power source.
- One or more sensors.
- Controller.

The controller is responsible for performing tasks such as processing data and controlling the functionality of other components of the sensor node. Microcontroller is the most commonly used controller; some other alternatives for microcontroller are a general-purpose desktop microprocessor, digital signal processors, FPGAs, and ASICs. A microcontroller forms the heart of many embedded systems which include sensor nodes because of its low cost, flexibility while connecting to other devices, ease of programming, and efficient power consumption. A general-purpose microprocessor has higher power consumption than a microcontroller; therefore, it is often not considered a suitable choice for a sensor node. Digital signal processors may be chosen for broadband wireless communication applications.

- Transceiver.

Sensor nodes basically use the ISM band that provides free radio, spectrum allocation, and global availability. The possible choices for wireless transmission media varies between radio frequency (RF), optical communication (laser), and infrared. Even though lasers require less energy, they require line-of-sight for communicating between devices and are sensitive to atmospheric conditions. Infrared, such as lasers, does not require an antenna and has limited broadcasting capacity. Radio frequency-based communication is the most relevant that fits most of the WSN applications. WSNs use license-free communication frequencies: 173, 433, 868, and 915 MHz and 2.4 GHz. The transceiver has combined functionality of transmitter and receiver but has no unique identifiers.

The operational states are as follows:

1. Transmitting.
2. Receiving.
3. Idle.
4. Sleep.

Recent advancement in technology of transceivers has built-in state machines which can perform few of the functions automatically.

12.7 Authentication for Sensor Data

One of the attacks against the wireless sensor network is to spoof a node and inject false sensor data into it. If the data is largely distinct from what is being reported by other nodes, it is possible to detect the issue using simple statistical analysis. However, if the false data closely resembles genuine data, detection can be very hard. One of the possible countermeasures for these types of attacks is the data origin authentication. Block cipher such as AES alone does not provide any data integrity protection. However, in order to accommodate data integrity, it has to be equipped with what is known as message authentication code (MAC).

In order to ensure proper security, we need to first know the nature of the data that is being processed/stored/transmitted and also how critical is the data for an individual or organization or environment.

Failing to safeguard personal data might bring about a situation where a person's bank account may be compromised, or it may even be a threat to his very life. Furthermore, regulations relating to data privacy and security are very much necessary for ensuring consumer-friendly business and services.

12.7.1 Authentication Using CBC Mode

Authentication can be accomplished with the help of CCMP (Counter Mode with Cipher Block Chaining Message Authentication Code Protocol). CCMP is based on AES and forms a part of the 802.11i standard for wireless networks. CCMP adopts

128 bit keys and a 48 bit initialization vector that can substantially reduce the chances of replay attacks. Any enhancement in CCMP in terms of privacy and security might require additional processing power.

12.7.2 Authenticated Encryption Applied to Sensor Data

DHT (digital humidity temperature) sensor which is a temperature humidity sensor outputs the timestamped temperature and humidity values in both Celsius and Fahrenheit as shown in Fig. 12.6.

ESP8266 has inbuilt Wi-Fi module which helps us to send and receive data. As ESP is a memory and battery constrained small chip, we need to look for such an algorithm that has operations that require less computations. AES uses functions like shifting and XORing of bit values which can be performed by using this constrained device. ESP has an inbuilt flash memory that helps it to boot, and it also has an integrated high-speed cache which supports to increase system performance and thereby lead efficient memory utilization which is very important for small integrated chips.

The ESP pin diagram is shown in Fig. 12.7 which has pins for both analog and digital data communication and pins for connecting vcc and ground.

The ESP32 (shown in Fig. 12.7) is used for connecting the DHT11 sensor. The ESP32 will run the program in the figure which generates timestamped data shown in the figure. Here the DHT sensor has three pins which are connected to the ESP through ESP digital pin 4 ground and voltage.

Figure 12.8 shows the snapshot of embedded c program that is run on ESP32. Here we can see two parts the setup part that initializes some variables and also prints the heading for clarity. Second is the loop function which runs until the ESP32 I status is on or until some other program is loaded to it. The connection of ESP32 with DHT11 sensor is shown in Fig. 12.9.

This data has a value for timestamp, humidity, and temperature. This can be written to a file, and then that data file can be passed to the CBC encryption process which outputs a file having data in its encrypted form. Now the encrypted file is sent to the destination where the receiver can run the AES-CBC decryption algorithm to

DHT11 sensor output connected to Esp32

Timestamp	Status	humidity	Temperature C	F	Heat Index C	F
21:14:51.069	-> OK	41.0	28.7	83.7	28.4	83.1
21:14:53.124	-> OK	41.0	28.7	83.7	28.4	83.1
21:14:55.175	-> OK	41.0	28.7	83.7	28.4	83.1

Fig. 12.6 Output of DHT11 sensor

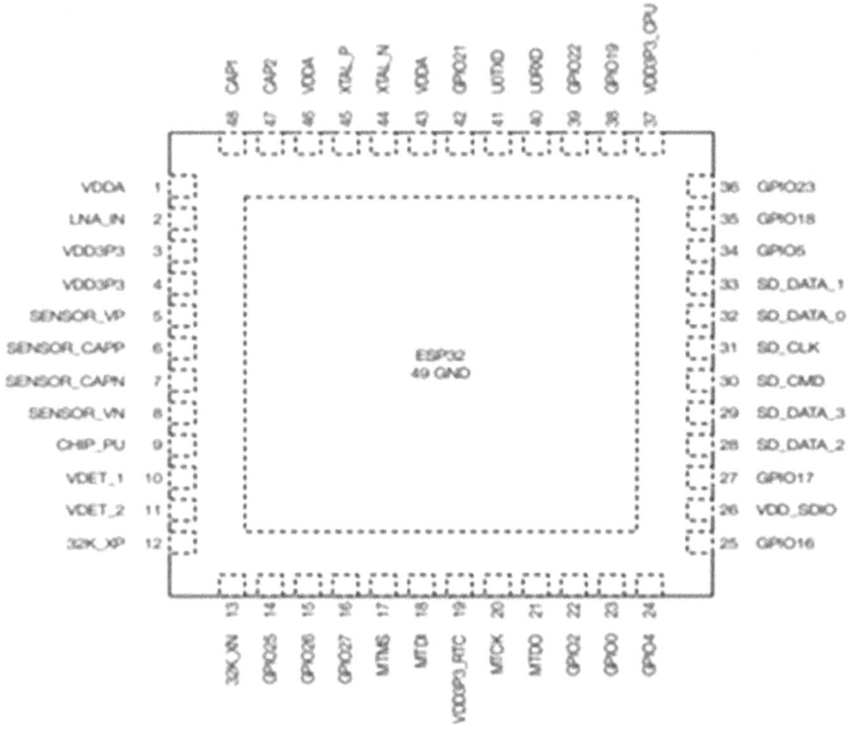

Fig. 12.7 ESP32 pin diagram

get back the original data. As CBC mode is used, the last block value will be the MAC for the whole file so if data is 1024 bits, last block data which is 128 bits should be added before sending/giving up a value of 1024 + 128 = 1152 bits of data. For the home automation, we can take into consideration all types of sensors like gas sensors, DHT11 for checking temperature, and buzzers in case of emergency situations. And all the data can be written to the file and can be encrypted using CBC-AES as explained above; the data will be encrypted and written to an output file.

12.8 Conclusion

Rijndael cipher is a symmetric block cipher that became AES after winning the competition to replace DES. It encrypts 128 bit blocks at a time with a 128/192/256 bit key. AES is safe to brute-forcing due to its key size. Chaotic variations of AES aim to increase the confusion and diffusion of the algorithm. This makes the algorithm less likely to be broken via confusion-diffusion and histogram analysis. ECC can be used to make an AES-ECC hybrid that can be used to transfer the AES block

```
#include "DHTesp.h"

#define DHTpin 4

DHTesp dht;

void setup()
{
  Serial.begin(115200);
  Serial.println();
  Serial.println("Status\tHumidity (%)\tTemperature (C)\t(F)\tHeatIndex (C)\t(F)");
  dht.setup(DHTpin, DHTesp::DHT11); //for DHT11 Connect DHT sensor to GPIO 17
}
void loop()
{
  delay(dht.getMinimumSamplingPeriod());

  float humidity = dht.getHumidity();
  float temperature = dht.getTemperature();

  Serial.print(dht.getStatusString());
  Serial.print("\t");
  Serial.print(humidity, 1);
  Serial.print("\t\t");
  Serial.print(temperature, 1);
  Serial.print("\t\t");
  Serial.print(dht.toFahrenheit(temperature), 1);
  Serial.print("\t\t");
  Serial.print(dht.computeHeatIndex(temperature, humidity, false), 1);
  Serial.print("\t\t");
  Serial.println(dht.computeHeatIndex(dht.toFahrenheit(temperature), humidity, true), 1);
}
```

Fig. 12.8 Program for DHT11 sensor

and key safely. AES is used in protecting data for storage, transmission (https and ssl), and in IOT (WPA2-AES). Attacks on AES include side-channel attacks and DFA attacks. By using authenticated encryption, sensitive data produced by sensors can be securely communicated between the senders and the receivers. As AES provides more security for securing sensor data, AES algorithm is used for more security, and for authentication AES with CBC is used. In this project, a real-time home automation system has been successfully implemented which is quite effective in terms of performance and technology. Home automation systems are not so popular yet. But, there is a very large probability that it will be the trend in the future. The proposed system has two advantages. First, using the IoT connectivity, we can monitor and access our smart home easily from anywhere, which will definitely prove to be energy efficient. Second, this provides a helping hand for the old age and differently abled person.

Fig. 12.9 Connecting DHT11 to ESP3

References

1. Pătru, I.-I., Carabaş, M., & Bărbulescu, M. (2016). Smart home IoT system, 2016 IEEE16th Edition, RoEdunet conference on September, 2016 – IEEE.
2. Madhumala, R. B., & Tiwari, H. (2020). Analysis of virtual machine placement and optimization using swarm intelligence algorithms. In A. Haldorai, A. Ramu, & S. Khan (Eds.), *Business intelligence for enterprise internet of things. EAI/Springer innovations in communication and computing*. Cham: Springer.
3. Govindraj, V., Sathiyanarayanan, M., & Abubakar, B. (2017). *Customary homes to smart homes using Internet of Things (IoT) and mobile application*. International Conference on Smart Technologies for Smart Nations on August, 2017 – IEEE.
4. https://www.electronicsforu.com/technology-trends/tech-focus/role-iot-home-automation
5. https://en.wikipedia.org/wiki/Internet_of_things
6. https://predictabledesigns.com/home-automation-with-an-arduino-a-basic-tutorial/
7. https://www.iotevolutionworld.com/m2m/articles/376816-history-smart-homes.htm
8. https://www.academia.edu/3549256/MODES_OF_OPERATION_OF_THE_AES_ALGORITHM
9. Yenuguvanilanka, J., & Elkeelany, O. (2008, April). *Performance evaluation of hardware models of Advanced Encryption Standard (AES) algorithm*. In Southeastcon, 2008. IEEE (pp. 222–225).
10. Madhumala, R. B., Tiwari, H., & Devaraj Verma, C. (2021). A reliable frame work for virtual machine selection in cloud Datacenter using particle swarm optimization. *International Journal of Mathematics and Computer Science, 16*, 677–685.
11. Pramstaller, N., Gurkaynak, F. K., Haene, S., Kaeslin, H., Felber, N., & Fichtner, W. (2004, September). *Towards an AES crypto-chip resistant to differential power analysis*. In Solid-state circuits conference, 2004. ESSCIRC 2004.
12. Jain, R., Jejurkar, R., Chopade, S., Vaidya, S., & Sanap, M. (2014). AES algorithm using 512 bit key implementation for secure communication. *International Journal of Innovative Research in Computer and Communication Engineering, 2*(3) 3516–3522.
13. Velayutham, R., & Manimegalai, D. Analysis of AES hardware and software implementation.
14. Kumar, A., Kumar, P. S., & Agarwal, R. (2019). *A face recognition method in the IoT for security appliances in smart homes, offices and cities*. 2019 3rd International Conference on Computing Methodologies and Communication (ICCMC). IEEE.

Index

M. Moh et al. (eds.), *Smart IoT for Research and Industry*, EAI/Springer Innovations in Communication and Computing,
https://doi.org/10.1007/978-3-030-71485-7

The manufacturer's authorised representative in the EU is Springer Nature Customer Service Centre GmbH, Europaplatz 3, 69115 Heidelberg, Germany. If you have any concerns regarding our products, please contact ProductSafety@springernature.com

Printed and bound by CPI Group (UK) Ltd, Croydon, CR0 4YY
30/09/2026
02225695-0002